ABIDIAN INDUSTRIAL INTERNET OF THINGS (IIoT)/ CONTINUOUS IMPROVEMENT DICTIONARY

A Working Index of Keywords Associated with the Industrial Internet of Things (IIoT), Lean, Six Sigma (6σ), Maintenance Excellence, Statistics, ISO Quality Management Systems, Theory of Constraints, and Total Productive Maintenance and Manufacturing (TPM)

Compiled and Edited by Mike Chambers

Austin Cove Books

Abidian Industrial Internet of Things (IIoT)/Continuous Improvement Dictionary: A Working Index of Keywords Associated with the Industrial Internet of Things (IIoT), Lean, Six Sigma (6σ), Maintenance Excellence, Statistics, ISO Quality Management Systems, Theory of Constraints, and Total Productive Maintenance and Manufacturing (TPM)

Copyright © 2017 Austin Cove Books
1937 W Palmetto Street #181
Florence, SC 29501
info@austincove.com

All Rights Reserved. Except as permitted under Sections 107 or 108 of the 1976 United States Copyright Act, no part of this publication may be reproduced, distributed, or transmitted in any form or by any means, or stored in a database or retrieval system, in whole or part, without the prior written permission of the publisher.

Trademarks: All trademarks are the property of their respective owners. The publisher and editor are not associated with and are not endorsing any product, vendor or third-party website mentioned in this book.

Limit of Liability/Disclaimer of Warranty: The publisher and editor make no representations or warranties with respect to the accuracy or completeness of the contents of this work and specifically disclaim all warranties, including without limitation warranties of fitness for a particular purpose. If professional research is required, the services of a competent professional person should be sought. Neither the publisher nor the editor shall be liable for damages arising herefrom.

Practical Manufacturing Series
ISBN-13: 978-1544803524
ISBN-10: 1544803524

Printed in the United States of America

There's no such thing as perfection. But, in striving for perfection, we can achieve excellence.

Vince Lombardi, Jr.
1913-1970

Table of Contents

Introduction 1
0-9 3
A 7
B 15
C 23
D 33
E 39
F 43
G 47
H 51
I 55
J 61
K 65
L 67
M 71
N 79
O 83
P 87
Q 95
R 97
S 101
T 111
U 117
V 121
W 125
X 129

Y	131
Z	133
Japanese Terms	135
About Abidian	141
About the Editor	143
Other Books by Mike Chambers	145
Index	147

Introduction

An obvious question is "Why the Industrial Internet of Things (IIoT) and Continuous Improvement (CI) in the same book?" On the surface, the answer may not appear intuitive. Experience has shown those interested in implementing IIoT are often doing so to reap the benefits of improved productivity. If the fundamental interest is in improved productivity, why not include IIoT and CI terms together in a single reference book? It also makes little sense to automate inefficiencies. Hence, Abidian's original Continuous Improvement Dictionary has been expanded to include both Continuous Improvement and Industrial Internet of Things terminology.

This working index of internet and improvement terminology was developed to provide a single, quick-reference for the Industrial Internet of Things (IIoT), Lean, Six Sigma (6σ), Maintenance Excellence, Theory of Constraints (TOC), ISO Quality Management Systems, and Total Productive Maintenance and Manufacturing (TPM) terms. As appropriate, cross-references have been provided to enhance understanding. Similarly, some entries include dual definitions or uses. When multiple interpretations are in wide use, an attempt has been made to incorporate each. The most common usage is shown first.

Also, included with the update is an extensive dictionary of commonly used Japanese improvement terms. It is important to note simple English translations often do not do justice to the intent behind some of the tools mastered in Japan.

In true Lean fashion, an effort has been made to be concise and straightforward in the definitions. Likewise, a common challenge in understanding and defining technical terms is distinguishing between in-house/local terminology and those terms in the public domain widely used by practitioners. An example is the acronym "TPM." To the purist, "TPM" is the Total Productive Maintenance methodology. However, some practitioners, recognizing the need to spread the tools across every aspect of manufacturing (and not just maintenance), refer to "TPM" as Total Productive Manufacturing. Appropriate explanations have been incorporated when this is the case.

There's change ... often mandated from above ... and then there's employee driven "excellent change." It is hoped the quick and easy definitions provided are helpful to both leadership and those employees in the trenches implementing positive change and their version of the Industrial Internet of Things (IIoT).

0-9

1.5 Sigma Shift
Belief (frequently open to discussion) that over time any process in control will shift from its target by a value up to 1.5 sigma

14 Management Principles (of Lean/The Toyota Way)
From Jeffrey Liker's *The Toyota Way*: 1) long-term philosophy, 2) create continuous flow, 3) use pull systems, 4) level the workload, 5) problem-solving culture, 6) standard work, 7) visual controls, 8) reliable/tested technology, 9) grow leaders who understand, 10) develop exceptional people and teams, 11) partner respect, 12) go see, 13) decide slowly by consensus and then rapidly implement, and 14) become a learning organization

14 Points
See Deming's 14 Points

2.4 GHz
Short-range wireless band used in wireless technologies like Bluetooth and Wi-Fi … See Bluetooth and Wi-Fi

3 Elements of Demand
Three (3) elements or drivers of customer demand and satisfaction are Quality, Cost, and Delivery … sometimes known as QCD

3 "G" Principles
Gemba (work area), *Genbutsu* (the actual object), and *Genjitsu* (reality/the facts) … a key to successful *kaizen*; i.e., improvement is 1) go to the work area, 2) work with the actual object, and 3) get the real facts … See *Gemba*, *Genbutsu*, *Genjitsu* and *Genchi Genbutsu*

THE ABIDIAN IIOT/CI DICTIONARY

3 Principles of Lean
1) Takt time, 2) One piece flow, and 3) Downstream pull from the customer … See One Piece Flow, Principles of Lean, Pull Production and Takt Time (T_T)

3 Step Change Process
1) Get ready, 2) Make it happen, and 3) Make it stick

3Ds
Dirty, dangerous, and difficult … sometimes dirty, dumb and dangerous (the first definition is usually preferred as it is more respectful) … Japanese version is known as the 3Ks (*kitanai* – dirty, *kiken* – dangerous, and *kitsui* – difficult)

3Ks
See 3Ds

3Ms
Three Japanese terms involving inefficiencies and variation: *Muda* (waste), *Mura* (irregular) and *Muri* (over burdening) … See *Muda*, *Mura*, *Muri* and Variation

3D Printing
See Additive Manufacturing

3P
See Production Preparation Process (3P)

3Rs
Three Reals: Reality, Real Place, and Real Thing

4th Industrial Revolution
Another term for the Industrial Internet of Things (IIoT) … earlier industrial revolutions were based on 1st) mechanized production (e.g., steam engine), 2nd) mass production and assembly lines (e.g., Ford's Model T) and 3rd) automation and computerized production (e.g., robots and PLCs) … See Industrial Internet of Things (IIoT) and Smart Manufacturing

4G
Fourth generation of cellular technology … provides greater data speeds versus the earlier 3G 3rd generation … See 5G

4Ms
Man/Woman, Machine, Materials, and Methods … See 5Ms of Production

4P Model
Continuous improvement model based on 1) Philosophy (long-term thinking), 2) Process (focus on flow and waste elimination), 3) People (deep respect for people and partners) and 4) Problem Solving (learning and root cause failure analysis)

THE ABIDIAN IIOT/CI DICTIONARY

5 GHz
Short-range wireless band used in wireless technologies like W-Fi ... 5 GHz provides faster data rates over a shorter distance than 2.4 GHz ... See 2.4 GHz and Wi-Fi

5 Whys
Root Cause Failure Analysis technique ... asking "why" repeatedly when a problem occurs to go beyond obvious symptoms to discover the root cause ... See Root Cause Failure Analysis (RCFA)

5G
5th generation technology for mobile cellular networks ... versus 4G, supports dramatically improved download speeds, greater user densities, and lower system latency (delay) ... See 4G

5Ms of Production
Man/Woman, Machine, Material, Method, and Measurement ... See 4Ms

5S
Five terms beginning with "S" describing practices conducive to visual control, lean production, and improved orderliness/organization ... the 5Ses are *Seiri* (Sort), *Seiton* (Store), *Seiso* (Shine), *Seiketsu* (Standardize) and *Shitsuke* (Sustain) ... Set in Order is sometimes substituted for Store ... some add a 6th S for Safety and refer to the process as 5S+1 or 6S

5S Audit
See 5S Scan

5S Cart
Supply station, typically on wheels, for 5S materials like tape, signage, labeling, and tools

5S Scan
Quick survey of an area for 5S compliance ... the term scan is often used instead of audit as it is usually considered less threatening

5S+1
5S tool with an added "S" for Safety ... See 5S

5W's and 1H
Tool to assist in problem identification and resolution ... Who? What? When? Where? Why? and How?

6 Sigma
See Six Sigma

6σ
See Six Sigma (6σ)

6S
See 5S and 5S+1

7 Flows
Flow of people, raw materials, subparts, final product, equipment, information, and engineering

7 Tools of SPC/SQC
Seven (7) Statistical Process/Quality Control (SPC/SQC) tools used by teams like Quality Circles: flow charts, histograms, pareto diagrams, scatter diagrams, cause and effect (fishbone) diagrams, control charts, and check sheets ... See Cause and Effect (Fishbone) Diagram, Check Sheet, Control Chart, Flow Chart, Histogram, Pareto Diagram, Quality Circle, Scatter Diagram, Statistical Process Control (SPC), and Statistical Quality Control (SQC)

7 Wastes
Original list of seven wastes identified by Toyota's *Taiichi Ohno*: transportation, inventory, motion, waiting, over processing, over production, and defects ... See *Taiichi Ohno* (1912-1990) and 8 Wastes

8 Wastes
Over Production, Waiting, Transportation, Over Processing, Inventory, Motion, Defects, and Skills/Resource Underutilization ... some combine Motion and Transportation and substitute an additional waste called "inefficiency" ... See 7 Wastes

802.11
See IEEE 802.11

80/20 Rule
See Pareto Principle

8D Corrective Action Process
Eight (8) step problem-solving process developed by the military to help ensure a corrective action is fully implemented: 1) Assemble cross-functional team, 2) Fully define the problem, 3) Implement and verify interim containment actions (i.e., temporary fixes), 4) Identify and verify the root cause, 5) Select and verify permanent corrective actions, 6) Implement and validate permanent corrective actions, 7) Prevent a reoccurrence of the problem/root cause, and 8) Recognize the efforts of the team

A3
Report (on a single A3-size sheet of paper) developed to present the thinking behind a problem, analysis, corrective action(s), and the associated improvement action plan ... see also A3 Problem Solving

A3 Problem Solving
Utilizing an A3 Report in a Plan-Do-Check-Act (PDCA) fashion where everything needed to guide a team through Root Cause Failure Analysis (RCFA) problem solving, reporting, and follow-up is available on a single sheet of paper ... see also A3, Plan-Do-Check-Act (PDCA), and Root Cause Failure Analysis (RCFA)

ABC
See Activity Based Costing (ABC)

Abnormality
Defect or something outside the standard(s) for a process ... See Abnormality Management and Anomaly Detection

Abnormality Management
Practice of readily identifying and immediately reacting to activities outside of the standard(s) for a process often through the use of visuals ... also known by the Japanese term *ijo kanri* ... See *ijo kanri* and Visual Workplace

AC
See Alternating Current (AC)

Accelerometer
Measures changes in gravitational acceleration that can be useful in measuring the acceleration, tilt, and vibration of an object

Acceptable Quality Level (AQL)
Quality level that, for the purposes of a sampling inspection, is the minimum limit of a satisfactory product or service; i.e., borderline "shippable"

Access Control
System that determines who or what can enter a system or facility/area when, where, and how ... a modern-day "door lock"

Access Control as a Service
Fee-based subscription for an access control service ... See Access Control

Access Point (AP)
Wi-Fi (expansion) node used to enhance signal strength and extend geographic coverage ... See Node and Wi-Fi

Accuracy
Measurement characteristic indicating how close an observed value is to the true value

Action Plan
Sequential, strategic steps to achieve results or an objective

Activity Based Costing (ABC)
Determining the actual cost of a product or service by tracking the cost of the specific activities that produce or provide the product or service

Activity Board
Visual tool sometimes associated with the Plan-Do-Check-Act (PDCA) methodology where planned activities are updated with actual activities and accomplishments in an effort to both drive and communicate continuous improvement progress ... See PDCA and Visual Workplace

Activity Network Diagram
Graphic used in planning a schedule for complex tasks and their related subtasks ... can depict scheduled completion times and adherence to the same ... See Project Evaluation and Review Technique (PERT) Chart

Activity Sheet
Identifies the team, objectives, current condition(s), problem(s) and any appropriate charts for a particular *kaizen* topic ... See Charter and *Kaizen*

Activity Tracker
Monitors and records physical/human data such as motion, temperature, presence, and heart rate

Actuator
Device that changes energy into motion or force ... converts electrical energy into mechanical energy

ADAS
See Advanced Driver Assistance System (ADAS)

ADC
See Analog-to-Digital Converter (ADC)

Additive Manufacturing
Process of making a three-dimensional object of virtually any shape from a three-dimensional model or electronic data source in which successive layers of material are added on top of one another ... sometimes referred to as 3D Printing

Addressability
Ability of a device or system to be targeted or found

Administrative/Service Wastes
Inefficiencies associated with an administrative, office, or service-oriented process ... sometimes referred to as people energy or people work wastes ... See 8 Wastes

Advanced Driver Assistance System (ADAS)
Vehicle system designed to enhance driver safety/performance

Advanced Equipment Standard (AES)
Electronic data encryption specification operating on a public/private key system

Advanced Message Queuing Protocol (AMQP)
Open-source standard for business messages and communications between applications or systems ... See Open-Source

Advanced Product Quality Planning (APQP)
QS-9000 process attempting to identify and solve opportunities before they become defects or problems affecting the customer ... See QS-9000

AES
See Advanced Equipment Standard (AES)

Aesthetics
Subjective sensory characteristics such as taste, sound, look and feel

Affinity Diagram
Brainstorming tool used to gather large quantities of information from many, often diverse, people (e.g., across shifts) ... ideas are visually displayed then categorized into similar columns ... columns are usually named indicating a general grouping of ideas

Agile Management
Iterative, highly flexible technique for managing the design and build processes for engineering, information technology, manufacturing, product development, and software development ... helps to enable rapid response to changing customer needs

THE ABIDIAN IIOT/CI DICTIONARY

AI
See Artificial Intelligence (AI)

Alexa
Voice-controlled intelligent personal assistant ... See Amazon Echo

Alpha (α)
Maximum probability a process or lot is unacceptable when, in fact, it is acceptable ... represented by the Greek letter α

AltBeacon
Open beacon specification ... See Beacon

Alternating Current (AC)
Electrical current that periodically changes direction

Amazon Echo
Smart speaker/intelligent personal assistant connected to the Amazon ecosystem

Amazon Web Services (AWS)
Amazon's remote computing services

Ambient Intelligence (AmI)
Capability to interpret and react to human activity ... some ambient intelligence systems also have the ability to learn

AmI
See Ambient Intelligence (AmI)

Amps/Amperes
Unit of electrical charge

AMQP
See Advanced Message Queuing Protocol (AMQP)

Analog
Signal capable of continuously changing over time between an "on" value and an "off" value

Analog-to-Digital Converter (ADC)
Converts an analog voltage or signal into a digital value representing the voltage/signal ... See Analog and Digital

Analysis of Means (ANOM)
Statistical procedure providing a graphical display of data for troubleshooting processes and analyzing the results of experimental designs

Analysis of Variance (ANOVA)
Statistical technique separating total data variation into meaningful component parts associated with specific sources of variation to test a hypothesis on the parameters of the model or to estimate variance components ... tool used to determine whether sample means are statistically equal by testing the variances of the samples ... Analysis of Variance can be used to compare two or more groups

Analyze
Process data is scrutinized for improvement opportunities ... causes affecting key input and output variables are verified ... third of the five stages in the Six Sigma DMAIC process ... See DMAIC and Six Sigma (6σ)

Anderson-Darling Normality Test
Used to determine if a group of data fits the standard normal distribution ... See Standard Normal Distribution

Andon
Japanese term (行灯 pronounced ahn-doan) for "lamp" ... a visual tool indicating the status of an operation at a single glance (e.g., flashing green, yellow, and red traffic light indicating the process is running, in need of attention, or the process has stopped) ... See Visual Workplace

Andon Cord
Cord or cable pulled by an employee to stop a process in the event of a defect ... See Andon

Android
Mobile operating system for touchscreen devices ... See Operating System (OS)

Android Wear
Open-source platform that extends Google's Android system to wearable devices ... See Open-Source and Wearable

ANN
See Artificial Neural Network (ANN)

Anode
Positive end of a diode ... See Diode

Anomaly Detection
Statistical technique to identify abnormalities ... See Abnormality and Abnormality Detection

ANOVA
See Analysis of Variance

Anthropometry
Study of the measurements and proportions of the human body ... science that defines physical measures of a person's size, form, and functional capacities ... See Ergonomics

AP
See Access Point (AP)

API
See Application Program Interface (API)

Application Program Interface (API)
Set of commands, routines, protocols, and programming tools for communicating between software programs or a software program and associated hardware

Appraisal Costs
Costs incurred to determine the degree of conformance to customer quality requirements

APQP
See Advanced Product Quality Planning (APQP)

Arduino
Open-source, inexpensive, and relatively easy to use hardware and software platform ... single-board microcontroller

Artificial Intelligence (AI)
Intelligence demonstrated by a machine or computer

AQL
See Acceptable Quality Level (AQL)

Arithmetic Mean
See Average and Mean

Arrow Diagram
Planning tool to graphically describe a sequence of events or activities (called nodes) and the interconnectivity between the same

Artificial Neural Network (ANN)
System of interconnected nodes (similar to the network of neurons found in the human brain) that is capable of learning ... See Neural Network (NN)

AS9100
International quality management standard for the aerospace industry published by the Society of Automotive Engineers and European Association of Aerospace Industries

Ashton, Kevin (1968-   )
Co-Founder of the Auto-ID Center at Massachusetts Institute of Technology who coined the term the "Internet of Things" ... under Ashton's direction, the Auto-ID Center also created a global standard for RFID devices and sensors ... See Radio Frequency Identification (RFID)

Assembly Line
From entry as raw material, parts are manipulated as they progress workstation by workstation toward a finished product at the end of the process

Asset Management
Systematic planning and control monitoring of a physical resource, asset, or piece of equipment throughout its life from conception to grave

Asset Tracking
Monitoring and reporting the movement and location of physical devices ... See Asset Management

Assignable Causes (of Variation)
Significant, identifiable changes in the relationships of man/woman, materials, methods, machines, measurement, and/or the environment

Attribute
Characteristic that can have only one value (e.g., 0 or 1, and yes or no)

Attribute Chart
Chart used in quality control to analyze a product or service to determine whether it is acceptable or defective

Attribute Data
Data that is counted rather than measured ... can have only one value ... See Attribute

Attribute Nominal Data
Form of discrete data characterized by labels and/or groupings (e.g., Department 1, Department 2 or Line 101, Line 102, etc.) ... See Discrete Data

Attribute Ordinal Data
Form of discrete data characterized by qualitative labels inherent to an object (e.g., excellent, good, fair, poor and handsome, average, homely) ... See Discrete Data

Automation
Use of electronics and control systems to operate equipment and processes often with little to no human interaction

Autonomation
Contraction of autonomous automation ... adding an element of human judgment to an otherwise automated process to assist in identifying unacceptable or subpar performance and quality ... sometimes referred to as automation with a "human touch" ... See *Jidoka*

Autonomous Maintenance
Technique recognizing the importance and involvement of an equipment operator in real-time monitoring, maintenance, and troubleshooting of their equipment and processes ... similar but more comprehensive (and possibly less controversial) term is Online Checks ... See Online Checks

Auto Time
Time when a process or equipment is running in automatic and an employee is not needed to operate the machine ... commonly applied to CNC machine cycles, oven cycles, rinse cycles, and wash cycles ... sometimes referred to as Automatic Time

Availability
Time equipment or a process is ready for use or operation ... in terms of OEE: a comparison of the potential operating time (typically, 24 hours a day, 7 days a week, and 365 days a year) and the time in which the equipment or process is actually making product/providing a service ... See Overall Equipment Effectiveness (OEE)

Average
Most common interpretation in continuous improvement is the arithmetic mean describing the central tendency of a data set ... calculated by summing all values and then dividing by the total number of values ... See Mean

AWS
See Amazon Web Services (AWS)

B

Backflow
Return of a product or service to an earlier step for reprocessing, rework, or repair ... See Hidden Factory

Backlog
Work that has not been completed and is not shown on the current/active work plan or schedule

Balanced
Available capacity is in sync with market demand

Balanced Scorecard
Business metric framework depicting strategic goals and operating performance indicators such as safety, customer satisfaction, sales, schedule compliance, internal production, employee utilization, and financial results

BAN
See Body Area Network (BAN)

Band
Range of (usually, telecommunication) frequencies

Bandwidth
Range within a given band of frequencies ... See Band

Bar Chart
See Bar Graph

Bar Graph
Graph using horizontal and vertical bars to represent the frequency of a given distribution; i.e., data grouped by category

Baseline
Performance level used as a standard of comparison for future performance

Batch and Queue
See Batch Operation

Batch Operation
Mass production approach to operations in which large lots (batches) of items are processed and moved to the next process often regardless of when/whether they are required ... objects typically wait in a queue until they can be further processed ... See Mass Production

Batch Size
Quantity of product worked in one process step ... See Batch Operation and One Piece Flow

Bathtub Curve
Plot (in the general shape of a bathtub) depicting the relationship of the life of many products and pieces of equipment versus the probable failure rate ... includes three curve portions: early or infant mortality (break-in), stable rate during normal operation and use, and wear out/fatigue ... See Weibull Distribution

Baud
Bits per second ... communications speed between devices or systems

Bayes Theorem
Statistical theorem relating conditional probabilities; i.e., computing the revised probability of an event occurring before another event when the events are dependent

Beacon
Device that broadcasts a usually unique identifier

Bell Curve
See Central Limit Theorem and Standard Normal Distribution

Benchmarking
Process of measuring products, services, and practices against companies or operations perceived as leaders ... intent is to determine how the leaders achieved their performance levels and then use the information to improve in-house performance

Benefit-to-Cost Ratio
Ratio comparison of the benefit(s) derived from an initiative to the cost(s) associated with the initiative

Best Practice
Demonstrated most productive, efficient, safest work method typically shared across an organization ... ideally, is the basis for standard work ... See Standard Work

Beta (β)
Maximum probability of a process or lot being acceptable when, in fact, it should be rejected ... represented by the Greek letter β

BI
See Business Intelligence (BI)

Bias (in Measurement)
Measurement characteristic referring to a systematic difference and/or pitfall in sampling ... an error leading to a difference between the average measurement result and the true, accepted value ... also known as a systematic error made in the selection of subjects from a given sampling ... See Sampling

Big Data
Sets of data so large it would be difficult to interact with them using conventional database tools

Binary
Only two states are possible (i.e., high or low, 0 or 1, and yes or no)

Binomial Distribution
Probability distribution involving a trial having only one of two possible outcomes (i.e., yes/no, pass/fail, heads/tails) ... one outcome has the probability p and the other the probability q where $p + q = 1$ ... the probability that the outcome p occurs r times in n trials is given by the binomial distribution

Bisertial Correlation Coefficient
Used with continuous variables when one variable is artificially reduced to two categories

Bit
Smallest piece of data or information a digital device can manage

Black Belt (BB) (Six Sigma)
Project team leader trained in the Six Sigma DMAIC methodology responsible for guiding improvement projects to completion ... as part of their training, the Black Belt will typically be required to demonstrate a proficiency in the use of advanced statistical techniques

Blast
Rapid process improvement ... See *Kaizen* Event

THE ABIDIAN IIOT/CI DICTIONARY

Block Diagram
Graphic depiction of the operations, interrelationships, and interdependencies of components in a system or process ... boxes or blocks (hence the name) represent the components with connecting lines between the blocks representing interfaces

Bluetooth
Short-range wireless technology standard ... depending on the version and class, Bluetooth can have speeds up to 50 Mbit/second and a range up to 240 meters (about 800 feet)

Bluetooth Low Energy
Bluetooth technology that optimizes battery life ... See Bluetooth

BMS
See Building Management System (BMS)

Body Area Network (BAN)
Wireless network of sensors and wearable devices ... See Sensor and Wearable

Boolean
Binary data type (i.e., only two states are possible) ... See Binary

Bootloader
Code that is executed on device power-up

Botnet
Network of malware-infected devices frequently used in attacks

Bottleneck
Point in an operation or process constraining (limiting) the flow of the overall process ... throughput is slowed because demand on the resource is greater than or equal to the bottleneck's capacity ... See Constraint, Monument, and Theory of Constraints (TOC)

Bounce
Software or physical device attempts to change its position or state but does not stay in that position and "bounces" back to its original position

Bowling Chart
Performance tracking sheet often used in startups and improving performance indicators/metrics ... See Key Performance Indicator (KPI)

Box Plot
Graph used to represent a data set when the data set contains a small number of values

BPE
See Budgeted Production Efficiency (BPE)

BPM
See Business Process Management (BPM)

BPMN
See Business Process Model and Notation (BPMN)

Brainstorming
Team-oriented technique to quickly generate ideas related to a subject or problem ... typically, team members offer their ideas and the ideas are immediately recorded, but not discussed or evaluated until after the brainstorming session has concluded

Breakdown
Failure in which a piece of equipment (or process) is unable to function as desired ... could be a complete failure or functional failure (e.g., noisy or will not achieve historical speeds or production levels)

Breakthrough
Paradigm shift after a chronic problem has been effectively solved to allow sustained performance at the next level of quality/production

Brick
Slang for an inoperable or defective device

Brillo
Internet of Things (IoT) operating system supporting Android devices, Bluetooth Low Energy, and Wi-Fi ... See Android Device, Bluetooth Low Energy, Operating System (OS), and Wi-Fi

Bring Your Own Device (BYOD)
Recognition employees and visitors often bring their personal devices into corporate environments ... See Corporate-Owned, Personally-Enabled (COPE)

Broker Server
"Middle" server that manages both incoming and outgoing data ... See Server

Brownfield
Construction additions or updates to an already existing facility

Browser
Software used to interact with and travel the internet

Bubble Diagram
Flow diagram utilizing bubbles, usually for material flow

Bucket Brigade
Employees slowly walk up and down a production line following a given product until they reach the end of the line or another employee ... if meeting another employee, the first employee starts tracking the second employee's product while the second employee starts slowly walking toward the front of the line where the process is repeated ... a benefit is in the event of a problem (i.e., line stoppage), all employees will ultimately wind up at the problem where they can help each other

Budgeted Production Efficiency (BPE)
Modification of the classic OEE calculation in which a plant or system's performance is compared to budgeted/planned production schedules, and line speeds ... See Overall Equipment Effectiveness (OEE)

Buffer Stock
Product accumulated because of uneven capacities and demands ... might also be product accumulated to protect the next internal process or end customer from starvation in the event of an abrupt increase in demand exceeding short-term production capacity ... a form of inventory and one of the 8 wastes ... See Inventory

Build-to-Order
Manufacturer builds products entirely to firm orders (rather than to a forecast or plan) ... also known as Make-to-Order (MTO)

Building Management System (BMS)
Monitors and controls a building's electrical and mechanical equipment such as lighting, HVAC, and access control

Built-in-Quality
Designing in quality and then producing the product in such a way it meets or exceeds customer specifications the first time ... See First Time Yield (FTY)

Bull-Whip Effect
Progressive magnification of demand upstream from the customer; i.e., each station in the value stream adds a safety factor ... See Demand Amplification

Business Intelligence (BI)
Results of data analysis used to assist in informed corporate decision making

Business Process Management (BPM)
Computer technology-based management philosophy promoting business effectiveness and efficiency while striving for innovation, flexibility, and integration

Business Process Model and Notation (BPMN)
Graphical representation of a process similar to a value stream map ... See Value Stream Mapping (VSM)

Business Value Added (BVA)
Process step where no value is added in the eyes of the customer (i.e., the customer is not willing to pay for it), but the step nonetheless adds organizational value (e.g., safety initiatives to reduce the potential for injury) … See Value Added Activities and Non-Value Added Activities

BusyBox
Collection of Unix utilities … See Unix

BYOD
See Bring Your Own Device (BYOD)

Byte
Typically, 8 bits … in most systems, a byte is a unit of 8 binary digits (i.e., bits) … See Binary and Bit

THE ABIDIAN IIOT/CI DICTIONARY

C

C of V
See Coefficient of Variation

C&E Diagram
See Cause and Effect (Fishbone) Diagram

C/O
See Changeover (C/O)

CA
See Certificate Authority (CA)

CAP
See Constrained Application Protocol (CAP)

Capability
Ability of a process to stay within specifications and on target ... See C_p and C_{pk}

Cargo Cult Science
Copying others without understanding ... may appear real, scientific, or religious but does not follow accepted practices ... See Fake Lean

Catchball
Give-and-take dialogue leading to shared objectives and consensus

Cathode
Negative end of a diode ... See Diode

Causal Flow Diagram (CFD)
Graphical representation of cause and effect relationships among major factors affecting a decision or problem

Cause and Effect (Fishbone) Diagram
Tool used to identify and organize in a structured format possible causes of a process or output problem ... sometimes called an *Ishikawa* Diagram (in tribute to its developer *Kaoru Ishikawa*) or a "fishbone" diagram (because it looks like the skeleton of a fish) ... See *Ishikawa, Kaoru* (1915-1989)

Cause and Effect Matrix
A template or tool, often spreadsheet-based, to assist in prioritizing the inputs to a conventional Cause and Effect (Fishbone) Diagram ... See Cause and Effect (Fishbone) Diagram

CBM
See Condition Based Monitoring (CBM)

c-Chart
Control chart used to assist in judging the quality of an item that could have one or more "countable" defects ... useful in determining if a process is in or out of control ... differs from a p-chart in that the c-chart accounts for the possibility of more than one nonconformity per unit

CC
See Creative Commons (CC)

CCPM
See Critical Chain Project Management (CCPM)

Cell
Processing steps for a product are located immediately adjacent to each other so parts and supplies can be processed in nearly continuous flow ... ideally, the cell will be in the shape of a "U" or "C" and typically flows counterclockwise ... See Cellular Manufacturing

Cellular Manufacturing
Equipment and workstations for a product family are arranged in close proximity (often in a U or C shape) to facilitate small lot, near-continuous flow production ... See Cell

Centaur Model
Hybrid combination of artificial intelligence and a human brain ... See Artificial Intelligence (AI)

Central Limit Theorem
States as the sample size increases, the shape of the distribution curve of the sample means will approach a normal distribution (i.e., a bell curve) ... See Standard Normal Distribution

Central Processing Unit (CPU)
Electronic "brains" and the main controller in a computer

Certificate Authority (CA)
Trusted entity that digitally verifies the internet identity of others

CFD
See Causal Flow Diagram

Chaku Chaku
Japanese term (着々 pronounced chah-kuu chah-kuu) literally translated as "load load" ... time and motion saving method of conducting one-piece flow in a cell where machines are arranged in the correct sequence and unload parts automatically ... See Cell and Cellular Manufacturing

Chance Variation
Random variation

Change Agent
Leader of a continuous improvement effort who has the will power and drive to initiate and sustain fundamental change

Changeover (C/O)
Process of switching from the production of one product, part, or service to another ... elapsed time between one product or service and the next one ... typically measured "good to good" (i.e., last good product to first good new product)

Charter
Contract between an organization and an improvement team documenting the purpose, objectives, resources, and boundaries of the team and the improvement

Chalk Circle Exercise
To increase awareness, Toyota's *Taiichi Ohno* would place an individual in a chalk circle and leave them ... on his return, he would ask "what have you seen?" ... if they had not seen a number of areas for potential improvement, he would leave them again ... this sequence was repeated until the person identified a sufficient number of opportunities ... See *Ohno, Taiichi* (1912-1990)

Chebyshev's Theorem
States the proportion of values from a data set falling within k standard deviation of the mean will be at least $1 - 1/k^2$, where k is a number greater than 1

Check Sheet
Standard work tool to help ensure all steps or items have been incorporated ... See Standard Work

Chi-Square Distribution
Probability distribution obtained from the values of $(n-1)s^2/\sigma^2$ when random samples are selected from a normally distributed population whose variation is σ^2

CI
See Continuous Improvement

Circuit
Path from a power supply, through a load, and back to the power supply

Client
Device that creates data for transmission or uses/retrieves data from another system ... also known as a publisher ... See Publisher

Cloud Computing
On user demand, storage and software applications are provided via the internet ... internet-based hosted services

CMMS
See Computerized Maintenance Management System (CMMS)

Code
Programming commands on a device or system

Coefficient of Determination
Measure of the variation of a dependent variable characterized by a regression line and independent variable ... the ratio of the explained variation and the total variation

Coefficient of Variation
Standard deviation (i.e., square root of the variation) divided by the mean ... expressed as a percentage ... See Variation

Cognitive IoT
Internet of Things (IoT) technology or systems that interact with humans through voice and text commands

Cognitive Radio
Capable of identifying and automatically switching to low-traffic radio frequencies

COGS
See Cost of Goods Sold (COGS)

Cold Chain
Refrigerated supply chain ... See Supply Chain

Common Cause Variation
Truly random sources of variability inherent to the process (rather than outside influences) ... See Variation

Companion Device
Wearable or similar device requiring a parent device (e.g., a smartphone) to operate or function

Complement of an Event
Outcomes in a sample space not in the outcomes of the event itself

Compound Event
Event consisting of two or more outcomes (or simple events)

Computerized Maintenance Management System (CMMS)
Computer-based system to assist in the management and administration of maintenance activities

Condition Based Maintenance (CBM)
See Condition Based Monitoring (CBM)

Condition Based Monitoring (CBM)
Process of basing maintenance, repairs, and operating practices (and subsequent intervention) on the monitored condition of a piece of equipment or system ... also known as Online Checks ... See Online Checks

Conditional Probability
Probability event B occurs after event A has already occurred

Conductor
Material that allows electricity to flow

Confidence Level
Probability a parameter will fall within a specified interval or range

Constrained Application Protocol (CAP)
Internet transfer protocol for low-power Internet of Things (IoT) devices

Constraint
Point in an operation or process limiting the overall flow of the process ... See Bottleneck and Theory of Constraints (TOC)

Consumer
See Customer

Continuous Improvement (CI)
Constant desire and effort to identify and implement ways to perform at ever higher levels

Continuous Initiatives
Classic approach to improvement: one initiative or program after another with limited efforts to sustain and subsequent declining performance ... frequently referred to as the flavor-of-the-month or fads

Continuous Data
Data measured along a continuous scale or continuum that can (theoretically) be infinitely divided

Continuous Flow
Producing and moving one item at a time through a series of processing steps as continuously/steady as possible with each step making only what is needed for the next step

Continuous Variable
See Continuous Data

Control
Identification and implementation of activities to standardize and incorporate new performance levels into daily activities so improvements might be sustained over the long term ... 5th and final stage in the Six Sigma DMAIC process ... See DMAIC and Six Sigma (6σ)

Control Chart
Time-series run chart having a centerline along with upper and lower statistical control limits ... quickly allows users to quickly determine if special cause variation has occurred or if a process is in control ... See Special Cause Variation

Controller
Electronic device that interfaces with and "controls" another device

Controls
Systematized methods and devices in place and actively used to prevent or detect failure modes or failure causes ... See *Poka Yoke* (Mistake Proofing)

Conveyance
See Transportation and the 8 Wastes

CONWIP
Constant-Work-in-Process (CONWIP): type of push-pull production control

COPE
See Corporate-Owned, Personally-Enabled (COPE)

COPQ
See Cost of Poor Quality (COPQ)

Corporate-Owned, Personally-Enabled (COPE)
Allows an end user to control most of a device while the corporation handles device security ... a compromise to BYOD ... See Bring Your Own Device (BYOD)

Corrective Action
Long-term root cause solution applied to a problem to prevent the problem or defect from occurring again ... See Problem Solving

Correlation
Statistical method to determine whether a relationship exists between two variables

Correlation Coefficient
Statistic indicating the strength and direction of a relationship between two variables ... See Correlation

Cost of Goods Sold (COGS)
Aggregate material and labor costs to produce a product

Cost of Poor Quality (COPQ)
Sum of costs incurred to prevent and detect problems, as well as to correct internal and external failures that escaped prevention efforts

Counter-Clockwise Flow
Typical flow of people and material in a work cell ... believed to have originated with machining equipment in that the chuck faced the right, making it easier for right-handed people to load from the right

Count of Items
Number of items ... typically expressed as "N" for a population and "n" for a sample

C_p
Indicator of the potential performance capability of a process ... measured as a ratio of the specification tolerance width to six standard deviations ... See Capability and Standard Deviation

C_{pk}
Indicator of the actual performance capability a process achieves ... measured as the minimum of the difference between the process average and the distance to the upper and lower spec limits divided by three standard deviations ... See Capability and Standard Deviation

CPPS
See Cyber-Physical Production System (CPPS)

CPS
See Cyber-Physical System (CPS)

CPU
See Central Processing Unit (CPU)

Creative Commons (CC)
Non-profit that helps manage retained copyright and royalty rights of work freely released to the public

Creative Commons Attribution – ShareAlike
Requires acknowledging and giving appropriate credit and then sharing in a similar manner ... See Creative Commons (CC)

Critical Failure
Involves damage or loss of function resulting in significant financial impact and/or jeopardizing safety, environmental compliance, quality, and basic operability ... See Critical Failure Mode

Critical to Quality/Customer (CTQ or CTC)
Product characteristic that satisfies a key customer or process quality requirement

Critical Value
Separates the critical region from the noncritical region in a hypothesis test

CRM
See Customer Relationship Manager (CRM)

Crosby, Philip B. (1926-2001)
Best known for developing the concept of Zero Defects and his 14 Steps to Quality Improvement ... key contributor to the concepts of Quality and Total Quality Management (TQM) ... See Zero Defects and Total Quality Management (TQM)

Cross-Platform
Capable of using different software (e.g., operating systems) on different computer types ... See Operating System (OS)

Crowdfunding
Venture capital raised from a large group of internet-based investors with each typically making a small contribution ... See Venture Capital

CryptoChip
Electronic component that manages cryptographic calculations

C_T
See Cycle Time (C_T)

CTQ or CTC
See Critical to Quality/Customer (CTQ or CTC)

Culture
See Lean Culture

Cumulative Distribution Function (CDF)
Result of adding the relative frequencies of a distribution ... the CDF is a function of the area underneath the Probability Density Function (PDF) ... See Probability Density Function (PDF)

Current
Rate of flow of an electrical charge through a circuit

Current State Map
Value Stream Map depicting a process or system as it currently exists and functions ... See Value Stream Mapping (VSM)

Customer
End person or entity (recipient) who orders and pays for goods or services

Customer Demand
Quantity of work units required by an end customer ... See Takt Time (T_T)

Customer Relationship Manager (CRM)
Software that "manages" customer information and communications

Customer Segmentation
Recognizing all customers are not equal and do not create equal value (profitability) for an organization, segmentation identifies those customers who generate the highest return (i.e., revenue, loyalty, purchase frequency, or lowest cost of doing business)

Cyber-Physical Production System (CPPS)
Computer control of manufacturing hardware ... See Cyber-Physical System (CPS)

Cyber-Physical System (CPS)
System that incorporates both computer (i.e., cyber) and mechanical (i.e., physical) characteristics and aspects

Cycle Time (C_T)
Time elapsed from the beginning of an operation to its completion; i.e., how often a part, product, step, or service is completed ... includes operating time plus the time required to prepare, load, and unload ... should be the fastest, repeatable time and not necessarily an average time

THE ABIDIAN IIOT/CI DICTIONARY

D

D of F
See Degrees of Freedom (D of F)

DAC
See Digital-to-Analog Converter (DAC)

Dashboard
Visual display of a group of summarized operational gauges, performance indicators, and metrics used to drive action/improvement

Data Center
Physical site providing computing and network services

Data-Driven Decision Management (DDDM)
Business decisions are made with and supported by verifiable data

Data Logger
Electronic device that transmits and possibly records sensor data

Data Mining
Analyzing large quantities of data to gain insight and to identify patterns

Data Scientist
Statistics and programming professional

Data Set
Collection of data values

Datakinesis
Situation where something (usually negative) occurs in the physical world as a result of some action in cyberspace

DC
See Direct Current (DC)

DDDM
See Data-Driven Decision Management (DDDM)

Debug
Troubleshoot (typically, a computer system or program)

Decision/Event Tree
Identify alternative scenarios and their component decisions and events

Defects
Events or outputs that do not meet the requirements of the customer and/or specifications of the process ... 1 of the 8 wastes ... See 8 Wastes and Non-Conformance

Defects Per Million Opportunities (DPMO)
Calculated by taking the total number of defects divided by the number of possible opportunities for defects, then multiplying by one million

Defects Per Unit (DPU)
Number of defects, problems, or issues per production or service unit

Define
For a Six Sigma project, reaching agreement on the scope, goals, and targets ... 1st of the five steps in the Six Sigma DMAIC process ... See DMAIC and Six Sigma (6σ)

Degrees of Freedom (D of F)
Number of values free to vary after a single statistic has been computed ... used when a distribution consists of a family of curves ... directions in which an object can move

De-Identification
Stripping data of personally identifiable data

Delivery Performance
Measures the ability to satisfy customers on shipped quantity and timeliness

Demand
See Customer Demand

Demand Amplification
Progressive magnification of demand upstream from the customer; i.e., each station in the value stream adds a safety factor ... See Bull-Whip Effect

Demand Response
Voluntary reduction of (usually) electrical consumption ... also known as load shedding ... See Load Shedding

Deming Cycle
See PDCA

Deming, W. Edwards (1900-1993)
American statistician and consultant best known for teaching the Japanese how to dramatically improve quality (and subsequent profitability) through statistical methods and a unique and insightful management philosophy ... See Deming's 14 Points and Total Quality Management (TQM)

Deming's 14 Points
1) Create constancy of purpose for improvement of product and service, 2) Adopt the new philosophy, 3) Cease dependence on mass inspection, 4) End the practice of awarding business on the basis of price tag alone, 5) Improve constantly and forever the system of production and service, 6) Institute training, 7) Adopt and institute leadership, 8) Drive out fear, 9) Break down barriers between staff areas, 10) Eliminate slogans, exhortations, and targets for the work force, 11) Eliminate numerical quotas for the workforce and numerical goals for people in management, 12) Remove barriers that rob people of pride of workmanship, 13) Encourage education and self-improvement for everyone, and 14) Take action to accomplish the transformation

Dependent Samples
Subjects are paired, matched, or related in some way

Descriptive Statistics
Branch of statistics consisting of the collection, organization, summarization, and presentation of data

Design for Manufacturing (DFM)
Design and development where the focus is on efficient manufacturing ... similar term is Design for Maintainability

Design for Six Sigma (DFSS)
Methodology seeking to avoid manufacturing and service problems from the very outset by utilizing Six Sigma statistical techniques in the design and development of products, systems, and services

Design of Experiment (DOE)
Systematic statistical study of many (often simultaneous) factors at multiple levels in order to determine their individual effect on the mean

Device
Piece of equipment, often smart with computing capabilities, that can interact with humans, computing systems, and other pieces of equipment

Device Attack
Exploiting a vulnerable device ... See Device

Devil's Advocacy
Seeking to prove, with the same or different evidence, a contrary or opposite view

DFM
See Design for Manufacturing (DFM)

DFSS
See Design for Six Sigma (DFSS)

Digital
Discrete values expressed in terms of 0 and 1

Digital-to-Analog Converter (DAC)
Circuit that converts a digital value or signal into an analog voltage ... See Analog and Digital

Diode
Electronic device that conducts electricity in only one direction ... an electronic "check valve"

Direct Current (DC)
Electric current that flows in only one direction, typically from positive to negative (ground)

Direct Memory Access (DMA)
Allows devices and peripherals to directly access a computing device's memory without CPU involvement

Discrete Data
Data that is counted ... binomial data that can have only one of two values (e.g., yes/no and pass/fail), nominal data characterized by labels/groupings (e.g., Department 1, Department 2), and attribute data characterized by a qualitative level inherent to an object (e.g., excellent, good, fair, and poor) ... See Attribute Data and Discrete Variable

Discrete Variable
A variable that can be "counted" ... See Discrete Data

Distribution
Pattern of numbers that is reproducible

Divergent Thinking
Process utilizing four commandments: 1) The more ideas the better, 2) Build one idea upon another, 3) Wacky ideas are okay, and 4) Don't (immediately) evaluate ideas

DIY
See Do It Yourself (DIY)

DL or D/L
See Downlink (DL or D/L) and Download (DL or D/L)

DMA
See Direct Memory Access (DMA)

DMAIC
Six Sigma methodology of Define, Measure, Analyze, Improve, and Control ... used to improve products and processes such that results are more predictable and approach defect free ... See Six Sigma (6σ)

DMADV
Acronym for the Six Sigma methodology of Define, Measure, Analyze, Design, and Verify ... typically used on existing processes

DNP3 Protocol
Open standard for the electric utility industry

Do It Yourself (DIY)
Building, modifying, or repairing something without the direct involvement of an expert

DOE
See Design of Experiment (DOE)

Downlink (DL or D/L)
Process of downloading or receiving data

Download (DL or D/L)
Copying data or information from one system to another

Downtime
Production time lost due to both planned and unplanned stops and outages ... includes equipment maintenance or failure, material and quality issues (including lack of raw materials), and training or staffing constraints

DPMO
See Defects Per Million Opportunities (DPMO)

DPU
See Defects Per Unit (DPU)

Drum-Buffer-Rope
Theory of Constraints (TOC)-based approach to production planning ... See Theory of Constraints (TOC)

THE ABIDIAN IIOT/CI DICTIONARY

E

Economic Order Quantity (EOQ)
Order quantity that optimizes both inventory levels and associated inventory holding and ordering costs

Economy of Scale
Belief there are cost benefits in increasing the size of an operation or customer base

ECRS
See Eliminate, Combine, Rearrange, and Simplify (ECRS)

Eddystone
Open beacon messaging format ... See Beacon

Edge Computing
Practice of locating applications and processing capabilities at the logical extremes of a network (rather than in the center)

Edge Gateway
Industrial Internet of Things (IIoT) protocol translator on the "edge" of the network

Edison
Platform for Internet of Things (IoT) devices and wearables ... See Wearable

EDR
See Electrodermal Response (EDR)

EEPROM
Electrically Erasable Programming Read Only Memory ... memory type that retains data without power

Effectiveness
Using the minimum amount of resources (people, materials, time, and space) with the least amount of waste to meet customer requirements

Efficiency
Minimum use of resources (including people, materials, time, and space)

Eight Wastes
See 8 Wastes

Electrodermal Response (EDR)
Changes in the electrical conductance of human skin as a result of activity or stimulation

Electrostatic Discharge (ESD)
Occurs if two objects with different electrical potentials come in contact with each other

Elements of Work
Value-added work, non-value-added work, and waste ... See Value Added Activities, Non-Value Added Activities, and Waste

Eliminate, Combine, Rearrange, and Simplify (ECRS)
Training-Within-Industry (TWI) approach to optimizing processes ... See Training Within Industry (TWI)

Embedded System
Computer system operating within a mechanical or electrical system

Embrace, Extend, and Extinguish
Corporate strategy to embrace open standards, expand the standards with additional capabilities, and ultimately eliminate the market viability of the open standards as a result of the enhanced capabilities

Emergency Stock
See Safety Stock

Empirical Data
Data that is acquired by direct observation and experimentation

Empirical Probability
Probability using frequency distributions based on observations to determine numerical probabilities of events

Empirical Rule
States when a distribution is normal (bell-shaped), approximately 68% of the data values fall within one standard deviation of the mean, 95% within two standard deviations, and 99.7% within three standard deviations ... See Bell Curve and Standard Normal Distribution

Enernet
Smart electrical grid

Enterprise Resource Planning (ERP)
Software to track data across a manufacturing facility ... includes accounting, production, purchasing, sales, and shipping ... See Material Requirements Planning (MRP)

EOQ
See Economic Order Quantity (EOQ)

Equipment Location Board
Visual display providing information on where equipment is located throughout a facility

ERP
See Enterprise Resource Planning (ERP)

Error Proofing
See *Poka Yoke*

Ergonomics
Practice of designing products, systems or processes to fit a job to a person ... also known as comfort design or functional design ... See Anthropometry

ESD
See Electrostatic Discharge (ESD)

Ethernet
System of protocols and hardware for connecting computers and computing devices to form a local area network

Executive Standard Work
Set of standard work practices for corporate executives as they visit a facility ... See Leader Standard Work (LSW)

Event
One or more outcomes of an experiment ... a *kaizen* might also be referred to as an improvement "event"

Excess Processing
See Over Processing

Excess WIP
Amount of work in process (WIP) above that called for in the standard work-based in-process inventory ... See Standard Work

Expert System
Software system that makes or evaluates decisions based on established logic or if/then rules (e.g., fault diagnosis)

Express Train
Sequence or path generated by following one specific part through a cell or value stream without any queue time

Extended Team Members
Sources of expertise and special skills outside of a core work group

Extensible Markup Language (XML)
Open standard that defines rules for encoding that is both human and machine-readable

Extensible Messaging and Presence Protocol (XMPP)
Communications protocol based on XML ... See XML

External Setup
Setup, changeover, or transition tasks that can be done while a process or machine is running ... See Internal Setup, Smart Changeover, and SMED

F

f Distribution
Sampling distribution of the variances when two independent samples are selected from two normally distributed populations in which the variances are equal and the variances s_1^2 and s_2^2 are compared as $s_1^2 \div s_2^2$

f-Test
Statistical test used to compare two variances or three or more means

Factor
Independent variable in an ANOVA test ... see Analysis of Variance (ANOVA)

FACTUAL
Problem-solving technique: Focus, Approach, Converge, Test, Understand, Apply, and Leverage

Failure
Equipment or process will no longer fully function (i.e., functional failure), perform at desired/historic levels, or meet a standard

Failure Effect
Consequences of a failure ... See Failure

Failure Mode
Manner in which a specific piece of equipment or process could fail to meet its requirements or intent

Failure Mode Analysis (FMA)
A Pareto-type analysis of failures using a frequency plot check sheet ... useful in highlighting problematic equipment and processes ... See Pareto Analysis

Failure Mode and Effect Analysis (FMEA)
Systematic method for identifying, assessing, and preventing design, product, process, and system failures ... could be a design-focused FMEA (i.e., a DFMEA) or process-focused FMEA (i.e., PFMEA) ... a comprehensive FMEA also includes the frequency and consequences of failure ... See Root Cause Failure Analysis (RCFA)

Failure Rate (λ)
Number of failures in a given timeframe ... represented by the Greek letter λ ... Failure Rate = 1/MTBF ... See Failure

Fake Lean
Looks like Lean but lacks understanding, thoroughness, and sustainment ... often performed by poorly trained employees at the request of leadership ... See Cargo Cult Science

FIFO
See First-In First-Out

Finished Goods
Completed products awaiting shipment to customer

Firmware
Programming or software that is written into the non-volatile read-only memory of a device ... often, automatically updated or capable of being updated by the end user

Firmware Over-the-Air (FOTA)
Automatically updating embedded software (firmware) over the internet without human involvement

First-In First-Out (FIFO)
Inventory management philosophy in which older inventory is processed first; i.e., first part to enter a process or storage area is the first to exit ... See Last-In First-Out (LIFO)

First Time Yield (FTY)
Percentage of conforming units or transactions produced on the first pass through an operation ... sometimes referred to as First Pass Yield

Fishbone Diagram
See Cause and Effect (Fishbone) Diagram

Five S
See 5S

Flavor-of-the-Month
See Continuous Initiatives

Flow
Movement of a product, service, material, or information through a value stream toward a customer

Flow Chart
Graphical representation of activities in a process that might include decisions, delays, movement and tasks ... See Statistical Process Control (SPC) and Statistical Quality Control (SQC)

Flow Production
Continuously moving product or information from one value-added step to the next ... See Just in Time (JIT) and One Piece Flow

Flow Shop
Shop floor arranged in the sequence of production flow

FMA
See Failure Mode Analysis (FMA)

FMEA
See Failure Mode and Effects Analysis (FMEA)

Focus Group
Group, typically composed of existing and potential customers, invited to discuss an existing or prospective product or service ... See Outside-In Thinking and Voice of the Customer (VOC)

Focused Improvement
One (1) of eight (8) typical Total Productive Maintenance/Manufacturing "Pillars" ... an eight-step, cross-functional, team-oriented approach to identify and eliminate process losses ... See Pillar, Total Productive Maintenance (TPM), and Total Productive Manufacturing (TPM)

Fog Computing
Decentralized networking approach where data and processing capabilities are placed throughout a network (typically, onsite) rather than in the "cloud" ... also known as fogging or a fog network

Fool Proofing
A disrespectful term typically not used in Lean Cultures ... better, less condescending terms are error or mistake proofing ... See *Poka Yoke*

Ford, Henry (1863-1947)
American industrialist and founder of the Ford Motor Company ... enhanced the concept of an assembly line and was a pioneer in mass production ... one of the originators of the concepts of waste elimination and standard work

FOTA
See Firmware Over-the-Air (FOTA)

Four Ms
See 4Ms

Franklin, Benjamin (1706-1790)
American statesperson and scientist whose philosophies regarding time, problem-solving, scientific experimentation and common sense are the backbone of many Lean initiatives

Frequency Distribution
Count of the number of times a value occurs in a sample or over several samples

Frequent Conveyance
Increasing the delivery frequency of parts while simultaneously reducing delivery quantities resulting in a lower inventory quantity at any given instant

Fuchsia
Android operating system for embedded devices

Future State Map
Value Stream Map depicting the desired process arrangement and metrics after improvements have been implemented; i.e., desired future to aspire to … some refer to this as the Ideal State Map and believe this is the most important of the two maps produced … See Current State Map and Value Stream Mapping (VSM)

G

Gage R & R
Measure of the percent of the total variation in the data that is taken up by measurement system error including both "repeatability" and "reproducibility" errors

Galvanic Response
See Electrodermal Response (EDR)

Gantt Chart
Planning method indicating the projected start and completion times for a project's scheduled activities ... See Project Evaluation and Review Technique (PERT) Chart

Gap Analysis
Evaluation of the current state versus some desired or optimal state ... typically incorporates actions required to move from the current to the desired state

Gateway
Link or portal between two computer systems or programs

Gates, Bill (1955-   )
Businessperson best known for co-founding Microsoft

Gauss, Johann Carl Friedrich/Frederick (1777-1855)
German mathematician credited with the development of the Gaussian Curve commonly known as the Standard Normal Distribution

GB
See Green Belt (GB)

Gemba
Japanese term (現場 pronounced gim-bah) ... where the action occurs: the "work area" or the "actual place" ... within manufacturing, the shop floor or production area ... see 3 "G" Principles and *Gemba* Walk

Gemba Walk
Lean management technique of walking the work area observing, understanding, asking questions, and learning ... See *Gemba* and Management by Wandering Around (MBWA)

Genbutsu
Japanese term (現物 pronounced gane-boot-sue) ... the actual product ... sometimes written *gembutsu* ... See *Genchi Genbutsu*

Genchi Genbutsu
Japanese term (現地現物 pronounced gen-chee gane-boot-sue) meaning to "go and see" ... go to the source or "*Gemba*" and learn/gain a deep personal understanding ... See *Gemba*

Genjitsu
Japanese term (現実 pronounced gane-jeet-sue) ... the facts (reality) ... See *Genchi Genbutsu*

Geofence
Virtual radius (or perimeter) around a real-world geographic (physical) area ... involves the use of location-aware devices (e.g., mobile smartphones) and location-based services

Geographic Information System (GIS)
System of hardware, software, and geographic data ... used by location-enabled devices

Gilbreth, Frank (1868-1924)
Time, motion, and human factors innovator ... developed the concept of "therbligs" used in motion studies ... See Therblig Analysis

GIS
See Geographic Information System (GIS)

Global Positioning System (GPS)
Satellite system used by "GPS receivers" to identify and navigate between locations

Goldratt, Eliyahu (1947-2011)
Author and management consultant best known for his development of the Theory of Constraints (TOC) toolset ... See Theory of Constraints (TOC)

Good Thinking, Good Products
Internal Toyota slogan to promote Toyota's in-house suggestion system

Goodness-of-Fit Test
Chi-square test determining if a frequency distribution fits a specific pattern

GPS
See Global Positioning System (GPS)

Graphical User Interface (GUI)
Boundary utilizing icons and other visuals between a human and a computer system … See Human Machine Interface (HMI)

Green Belt (GB)
Person trained in Lean Six Sigma/Six Sigma who is then expected to lead improvement projects within her organization under the guidance of a Black Belt … training is typically 80 hours (2 weeks) or less

Green Washing
Deceptive marketing practice of leading the public to believe a product or practice is good for the environment when in fact it is often harmful to the environment

Greenfield
New construction where everything is designed and constructed from a starting point of an empty "green field"

Ground
Point in an electrical circuit with zero potential electrical energy

GUI
See Graphical User Interface (GUI)

THE ABIDIAN IIOT/CI DICTIONARY

H

Hadoop
Java-based programming language used for handling large data sets ... See Big Data

HaLow
IEEE 802.11ah 900 MHz Wi-Fi protocol ... See IEEE 802.11 and Wi-Fi

Hanedashi
Japanese term (はね出し pronounced hahn-dah-she) for an auto-eject device that unloads a part allowing an operator to go from one machine or workstation to the next picking up and loading ... See *Chaku Chaku*

Hansei
Japanese term (反省 pronounced hahn-say) for critical self-reflection ... even after success, reflecting on how an experience could have been even better ... in Japan, *hansei* is an emotional and introspective experience with three elements: 1) The organization (or individual) must recognize there is a problem; i.e., a gap between expectations and achievement and subsequently be receptive to potentially negative feedback, 2) They must voluntarily take responsibility and feel deep regret and 3) They must commit to a specific course of action to improve

Haptic Technology
Science involving touch, feel, and tactile sensations ... also known as haptics

Haptics
See Haptic Technology

Harada Method
Technique to enhance employee/leadership development

Hawthorne Effect
Process improves merely because it is being observed ... often reverts to original performance levels when observations are stopped

HAZOP
HAZard and OPerability study intended to proactively identify needed equipment or process modifications to avoid safety or environmental incidents because of equipment or human failure

Heijunka
Japanese term (平準化 pronounced high-joon-kah) for load leveling the type and quantity of production over a fixed period of time ... redistributes work in proportion to customer demand factoring in volume, variety, and mix ... also known as smoothing

Heijunka Box
Physical device controlling production mix and volume through a *kanban* at fixed intervals of time ... the "box" is divided into slots containing *kanban* cards representing pitch increments ... See *Heijunka, Kanban*, and Pitch

HEM
See Home Energy Management (HEM)

Hertzian Space
Hidden electromagnetic environment created by ever increasing numbers of wireless devices

Hexadecimal
Numeric system with 16 values (usually, 0-9 and A-F) ... also known as Base 16 or Hex

Hidden Factory
Workaround or re-work loops outside the intended flow of production that are necessary to correct errors and defects

HIPPO
Acronym for HIghest Paid Person's Opinion

Histogram
Graphical tool (Bar Chart) depicting the shape or distribution of data by showing how often different values occur ... bars of proportional height/length represent the frequency of distribution ... sometimes referred to as a Frequency Diagram ... See Pareto Analysis

HMI
See Human Machine Interface (HMI)

Home Energy Management (HEM)
Smart monitoring and control of residential energy (i.e., electricity, water, solar, and natural gas)

Hoshin
Japanese term (方針 pronounced hoe-sheen) for strategic and tactical planning

Hoshin Kanri
Japanese term (方針管理 pronounced hoe-sheen kahn-ree) for cascading, top-down strategic/tactical planning ... also known as policy deployment, strategic planning, *hoshin* planning, or simply *hoshin* ... See *Hoshin* and Strategic Planning

Host
Computing system(s) that provides services or resources to a network

House of Quality
Tool used in the Voice of the Customer process that helps to quantitatively consider customer needs, relative importance, competitive position, measures, conflicts, and relationships ... See Voice of the Customer (VOC)

Hub
Computer node that is usually at the "center" of a system ... often contains software that monitors and controls a network or system ... the center of a wheel

Human Machine Interface (HMI)
Boundary between a person and a machine or computer

Human Optimization
Optimizing the performance and motion of people with the assistance of wearables ... See Wearable

HVAC
Acronym for Heating, Ventilation, and Air Conditioning

Hypergeometric Distribution
Variable distribution having two outcomes when sampling is done without replacement

Hypothesis Tests
Decision-making process for evaluating claims about a given process or system

THE ABIDIAN IIOT/CI DICTIONARY

I

I/O
See Input/Output (I/O or IO)

IaaS
See Infrastructure as a Service (IaaS)

iBeacon
Technology capable of geographically locating both iOS and Android devices ... location services could be indoors or outdoors and could trigger events such as providing information or instructions

IC
See Integrated Circuit (IC)

ICS
See Industrial Control System (ICS)

IDE
See Integrated Development Environment (IDE)

Ideal State Map
See Future State Map

IEC
Acronym for International Electrotechnical Commission (IEC)

IEEE
Acronym for Institute of Electrical and Electronic Engineers (IEEE)

IEEE 802.11
Standard for wireless network communications

IFTTT
If This, Then That ... system where conditional statements trigger action

IGES
See Initial Graphics Exchange Specification (IGES)

IGMP
See Internet Group Management Protocol (IGMP)

IIoT
See Industrial Internet of Things (IIoT)

IISF
See Industrial Internet Security Framework (IISF)

Ijo kanri
Japanese term (異常管理 pronounced e-joe kahn-ree) for abnormality management ... See Abnormality Management

Improve
Pilot and implement solutions to better a process, service, or product ... 4th step in the Six Sigma DMAIC process ... See Six Sigma (6σ)

Indoor Positioning System (IPS)
System to location people, objects, and devices within a building

Industrial Control System (ICS)
Computing system that monitors and controls a manufacturing or industrial process

Industrial Internet of Things (IIoT)
Science of interfacing industrial machines to enable the machines to communicate and exchange data ... interconnection of automated systems and information technology ... See Information Technology (IT) and Smart Manufacturing

Industrial Internet Security Framework (IISF)
Security protocol covering industrial safety, reliability, resilience, security, and privacy

Industry 4.0
Strategy that began in Germany that promotes cyber-physical systems monitoring physical manufacturing processes while making decentralized decisions ... another term for the 4th Industrial Revolution and the Industrial Internet of Things (IIoT) ... See 4th Industrial Revolution and Industrial Internet of Things (IIoT)

Infant Mortality
Relatively high probability of failure (versus normal run-time failures) for new equipment and or repaired equipment immediately after an attempt has been made to start it

Inferential Statistics
Branch of statistics consisting of generalizing samples and populations, performing hypothesis testing, determining variable relationships, and making statistics-based predictions

Information Flow
Flow of communications, including supporting data and work instructions, along a value stream ... See Material Flow and Value Stream Mapping (VSM)

Information Technology (IT)
Study and use of computing and telecommunications systems

Infrastructure as a Service (IaaS)
Fee-based subscription for virtual or cloud-based computing resources

Initial Graphics Exchange Specification (IGES)
File format standard used to transfer files between computer-aided design applications

Input/Output (I/O or IO)
Received data (i.e., input) and transmitted data (i.e., output) ... sometimes is in reference to the system or device ports that receive and transmit

Input
Received signal or data

Input X's
See X Data/Values

Inspection
Determining the condition of a piece of equipment, process, or product typically as it compares to some standard ... often reactive/after-the-fact and, as a result, does not directly prevent the defect from occurring

Installer
Computer application that extracts and installs software

Insulator
Material that inhibits or slows the flow the electrical current

Insurance Telematics
Vehicle tracking technology where a tracking device to monitor driving habits is offered by an insurance company in return for lower insurance rates

THE ABIDIAN IIOT/CI DICTIONARY

Integrated Circuit (IC)
Semiconductor-based electronic circuit ... See Semiconductor

Intelligent Device
Instrument, machine, or piece of equipment with integrated computing capabilities ... also known as a smart device ... See Smart Device

Intelligent Transport System (ITS)
Smart system installed in vehicles to improve mobility and safety

Interaction
Condition in which the effect of the level of a factor on a response is different for different levels of a second factor

Interaction Effect
Effect of two or more variables on each other in a two-way ANOVA ... See Analysis of Variance (ANOVA)

Interdependence Test
Chi-square test used to validate the interdependence of two variables when data are tabulated in terms of frequency

Interdependent Variable
Correlation and regression analysis variable that can be controlled or manipulated ... See Correlation and Regression

Integrated Development Environment (IDE)
Software/code editor

Internal Setup
Setup, changeover, or transition tasks that cannot be done while the process or machine is running ... See Smart Changeover and SMED

International Telecommunications Union (ITU)
United Nations agency responsible for information and communication technologies

Internet
Global system of interconnected academic, business, government, private, and public computer networks and sites ... See Intranet

Internet Group Management Protocol (IGMP)
Internet protocol – based technology for group communications ... See Internet Protocol (IP)

Internet of Things (IoT)
Connecting devices (usually smart or with embedded electronics) through an internet/intranet to enable the devices to communicate and exchange data ... consumer version of the Industrial Internet of Things (IIoT)

Internet Protocol (IP)
Principal communications protocol for transmitting data across the internet

Interrelationship Digraph
Graphically displays the sequential interrelated factors involved in a complex, multi-variable problem

Intranet
Private, typically in-house, network that is usually accessible internally to a single organization ... See Internet

Inventory
Raw material, supplies, work in process, or finished goods in excess of current needs ... one of the 8 wastes ... See 8 Wastes

Inventory Turns
How quickly material passes through a facility or value stream ... can be approximated by dividing the annual Cost of Goods Sold (COGS) by the average inventory value ... Annual Inventory (Stock) Turns is calculated by dividing the annual cost of goods sold by the average on-hand inventory (at plant cost) ... total inventory includes raw materials, work in process, and finished goods ... plant cost includes material, labor, and plant overhead

IO
Input/Output (I/O or IO)

IoT
See Internet of Things (IoT)

IP
See Internet Protocol (IP)

IP Address
Unique internet protocol identifier assigned to devices connected to an intranet or the internet

IPS
See Indoor Positioning System (IPS)

iPv4
Internet Protocol Version 4 ... used to route packet-switched internet traffic ... See IP Address and Packet Switching

iPv6
Internet Protocol Version 6 ... used to route packet-switched internet traffic ... uses 128-bit (theoretically 3.4×10^{38}) addresses to deal with the exhaustion of the limited 32-bit iPv4 addresses (4.3×10^{9}) ... See IP Address, iPv4, and Packet Switching

Ishikawa Diagram
See Cause and Effect (Fishbone) Diagram

Ishikawa, Kaoru (1915-1989) (石川馨 pronounced kah-oh-ruu ee-she-kah-wah)
Japanese quality professional best known for his development of Quality Circles and the Cause and Effect (sometimes known as the *Ishikawa*) Diagram ... See Cause and Effect (Fishbone) Diagram and Quality Circle

ISO
International Organization for Standardization ... independent, non-governmental organization with a membership of over 160 national standards organizations

ISO 13053
Quantitative methods in process improvement -- Six Sigma -- Part 1: DMAIC methodology and Part 2: Tools and Techniques ... See ISO and Six Sigma

ISO 14001
Environmental Management Systems -- Requirements with Guidance for Use ... See ISO

ISO 17258
Statistical methods -- Six Sigma -- Basic Criteria Underlying Benchmarking for Six Sigma in Organizations ... See ISO and Six Sigma

ISO 18404
Quantitative methods in process improvement -- Six Sigma -- Competencies for key personnel and their organizations in relation to Six Sigma and Lean implementation ... See ISO, Lean and Six Sigma

ISO 9001
Quality Management Systems – Requirements ... See ISO and Quality Management System (QMS)

ISO/IEC 29161
Information technology -- Data structure -- Unique Identification for the Internet of Things ... See IEC, ISO, and Internet of Things (IoT)

IT
See Information Technology (IT)

ITS
See Intelligent Transport System (ITS)

ITU
See International Telecommunications Union (ITU)

J

Japanese Terms
See section on Japanese Terms

JavaScript Object Notation (JSON)
Alternative to XML ... text-based and readable by humans format that uses name and object pairs to organize data

JCAHO
See Joint Committee for the Accreditation of Healthcare Organizations (JCAHO)

JerryScript
Lightweight JavaScript for low-energy memory-constrained devices like microcontrollers

JI
See Job Instruction (JI)

Jidoka
Japanese term (自働化 pronounced jee-doe-kah) for providing machines and operators the ability to 1) detect when an abnormal condition has occurred and to 2) immediately stop the work ... quality at the source ... autonomation: adding human judgment to an otherwise automatic process ... See Autonomation

Jishuken
Japanese term (自習 pronounced jee-shuu-ken) for self-study ... may also refer to a combination training-improvement event led by management ... See *Kaizen*

JIT
See Just in Time (JIT)

JM
See Job Methods (JM)

Job Instruction (JI)
Training Within Industry (TWI) technique to ensure employees can perform a task correctly, safely, and consistently ... See Standard Work and Training Within Industry (TWI)

Job Methods (JM)
Training Within Industry (TWI) technique teaching employees how to objectively evaluate their work tasks with an eye toward continuous improvement ... See Standard Work and Training Within Industry (TWI)

Job Relations (JR)
Training Within Industry (TWI) technique teaching leadership how to effectively and fairly interact with employees ... See Training Within Industry (TWI)

Job Shop
Manufacturing facility, often small to mid-size, that typically handles small batch jobs and is arranged by skill or process (i.e., welding, machining, and painting) rather than in process sequence

Jonah
An expert in Theory of Constraints (TOC) ... from the character of the same name in Eliyahu Goldratt's *The Goal* ... See Goldratt, Eliyahu (1947-2011) and Theory of Constraints (TOC)

Jones, Daniel T.
With James P. Womack, co-authored *The Machine That Changed the World* which coined the term "Lean Production" ... Jones was later the founder and chairman of the United Kingdom's Lean Enterprise Academy

Joint Committee for the Accreditation of Healthcare Organizations (JCAHO)
Sets standards, evaluates and accredits healthcare organizations in the United States

JR
See Job Relations (JR)

JSON
See JavaScript Object Notation (JSON)

Juran, Joseph M. (1904-2008)
20th-century quality management consultant ... went to Japan in 1954 and worked with Japanese senior and middle managers on quality management and improvement

Juran Trilogy
Three quality management processes developed by Joseph Juran: 1) Quality planning, 2) Quality control, and 3) Quality improvement

JUSE
See Union of Japanese Scientists and Engineers

Just in Sequence
Delivery of materials, work-in-process or finished product in the exact sequence it is needed

Just in Time (JIT)
Inventory management strategy in which materials, supplies, work in process, and finished goods are delivered immediately before and where they are needed ... makes and delivers what is needed, precisely when it is needed, in just the right quantity and quality ... intent is to avoid inventory build-up (one of the 8 wastes) ... see Inventory and 8 Wastes

Just in Time Training
Providing training immediately before it's needed and used to avoid acquired skills and knowledge being forgotten due to a lag between training and use

THE ABIDIAN IIOT/CI DICTIONARY

K

Kaiaku
Japanese term (改悪 pronounced kye-ah-kuu) for change for the worse ... a bad change ... is the opposite of *kaizen* ... See *Kaizen*

Kaikaku
Japanese term (改革 pronounced kye-kah-kuu) for dramatic, radical, revolutionary, and quick improvement of a process or value stream ... See Value Stream

Kaizen
Japanese term (改善 pronounced kye-zen) for continuous improvement of a process or value stream ... team-oriented effort to improve ... some regard as small, continuous improvements ... also known as a team-oriented, ongoing philosophy of constantly evaluating and improving ... may also be referred to as a blitz or workout ... See Charter and *Kaizen* Event

***Kaizen* Event**
Application of the *kaizen* continuous evaluation and improvement philosophy ... some refer to as a *kaizen* improvement event, pit stop, or blitz ... intent is the ultimate elimination of waste via daily, continuous improvements ... emphasis is on swift opportunity improvement by a dedicated team ... duration is hours to 5-days maximum ... See Continuous Improvement and *Kaizen*

Kalman Filter
In the presence of uncertain and dynamic information, a Kalman Filter facilitates an educated guess as to what the system will do next

***Kamishibai* Board**
Japanese term (紙芝居 pronounced kah-me-she-bye) for a paper drama or picture show ... within Lean, a *kamishibai* board is a visual task board ... See Visual Workplace

Kanban
Japanese term (看板 pronounced kahn-bahn) literally translated as "signal" ... signaling device that identifies a need and then gives authorization and instructions for the movement, production, or withdrawal of items in a pull system ... signal might be a floor marking, card, sheet, or light ... See Just in Time (JIT)

Kano Analysis
Method to better understand what value customers place on features of a product or service; i.e., what merely meets needs versus what truly excites and delights ... See Voice of the Customer (VOC)

Kano Model
Defines the hierarchy of customer needs: basic, performance, and excitement ... See Kano Analysis and Voice of the Customer (VOC)

Karoshi
Japanese term (過労死 pronounced kah-roe-she) for death from overwork ... See *Muri*

Kata
Japanese term (型 pronounced kah-tah) meaning form or routine ... within continuous improvement, there are Improvement *Katas* and Coaching *Katas*

Katashiki Card
Japanese term (型式 pronounced kah-tah-she-kee) literally translated as vehicle or model ... a document identifying all the options for a product model

Key Performance Indicator (KPI)
A select bottom-line-oriented measurement comparing performance against important business targets, goals, and objectives ... intent is to drive improvement

Kingman's Formula
Approximates the average waiting time in a que as a function of utilization, variability, and service time

Kitting
Grouping component parts for ease of use ... done properly, a visual management tool in that the user can readily see what has been consumed and what has not ... See Visual Workplace

KPI
See Key Performance Indicator (KPI)

Kruskal-Wallis Test
Nonparametric test comparing three or more means

L

Labor Linearity
Philosophy for flexibly staffing a production process or cell so the number of operators increases or decreases with production volume (i.e., the amount of human effort per part remains level (linear) as production volume changes)

LAMBA
Mnemonic for Look, Ask, Model, Discuss, and Act ... See PDCA

LAN
See Local Area Network (LAN)

Last-In First-Out (LIFO)
Early orders, products, or services are the last out to the customer ... See First-In First-Out (FIFO)

LCC
See Life Cycle Cost (LCC)

LCL
See Lower Control Limit (LCL)

Lead Time
Elapsed time from when a customer places an order to the customer receiving the order ... includes order processing time, cycle time, and the time associated with any waiting

Leader Standard Work (LSW)
Structured approach for team leads, supervisors, and managers to create a safe and efficient work environment that has a culture of continuous improvement ... See Executive Standard Work and Standard Work

LED
See Light Emitting Diode (LED)

Lean
Systematic, logical method of identifying and eliminating inefficiencies and non-value-added activities using continuous assessment and improvement to achieve improved effectiveness in response to customer demand ... Western world terminology to characterize Japanese productivity improvement techniques ... originally popularized by the book *The Machine That Changed the World* ... paradigm based on the elimination of waste while simultaneously maximizing flow ... See Principles of Lean

Lean Culture
Engaged work environment characterized by deep respect for people and partners, a long-term philosophy driven by continuous improvement, consistent elimination of waste and inefficiencies, and problem identification, solution, and elimination ... immeasurable, undocumented norms and guiding principles an organization embraces with its daily actions

Lean Enterprise
Continuing agreement (typically among facilities and sometimes companies) sharing a value stream to correctly specify value from the eyes of the end customer, remove waste, and make those actions which do create value occur in a continuous flow as pulled by the customer ... an organization fully understanding, communicating, implementing, and sustaining Lean concepts seamlessly throughout

Lean Principles
See Principles of Lean

Left Hand/Right Hand Analysis
Watching an operator's hands as she performs a process step ... intent is to reduce motion, one of the 8 Wastes

Level of Significance
Maximum probability of committing a Type I Error (i.e., error that occurs if one rejects the null hypothesis when it is true)

Level Production
See *Heijunka*

Level Scheduling
Averaging both volume and mix to create a more consistent flow and production level

Level Selling
Characterized by removal of sales discounts, incentives, and promotions ... intent is to remove demand spikes caused by fluctuations in the way a customer orders

Library
Collection of computing programming resources used in the development of software applications

Life Cycle Cost (LCC)
All costs associated with the acquisition, ownership, operation, maintenance, and ultimate disposal of a system or piece of equipment over its full life

Li-Fi
See Light Fidelity (Li-Fi)

LIFO
See Last-In First-Out (LIFO)

Light Emitting Diode (LED)
Two-lead semiconductor that generates light from its electroluminescence (activation)

Light Fidelity (Li-Fi)
Wireless technology based on visible light (i.e., infrared or near-ultraviolet) communications rather than radio waves ... See Visible Light Communication (VLC)

Line Balancing
Syncing takt time to cycle times and staffing ... evenly distributing work to each workstation and worker ... See Cycle Time (C_T), Takt Time (T_T), and *Yamazumi* Chart

Linux
Unix-like open-source operating system ... See Unix

LiPo Battery
Rechargeable lithium-ion polymer battery

Little's Law
Over the long term, the average number of customers in a stable system (L) is equal to the average arrival rate (λ) multiplied by the average time a customer spends in the system (W) .... expressed mathematically as $L = \lambda W$ ... within manufacturing, relationship between work-in-process, lead time, and the average time between parts ... See Lead Time and Work in Process (WIP)

Load
Electromechanical device that consumes energy

Load Shedding
Voluntary reduction of (usually) electrical consumption ... also known as demand response ... See Demand Response

Local Area Network (LAN)
Interconnected devices normally in close proximity that can communicate without the use of outside leased telecommunications systems … See Wide Area Network (WAN)

Long Range (LoRa)
Radio modulation format

Long Term Evolution (LTE)
Type of cellular network offering comparatively better data transfer speeds … sometimes known as 4G … See 4G

LoRa
See Long Range (LoRa)

Lot Size
Number of parts or products produced during a given production run, setup, or tooling configuration

Lower Control Limit (LCL)
Minimum acceptable value of a mean or range of a sample

LSW
See Leader Standard Work (LSW)

LTE
See Long Term Evolution (LTE)

M

M2M
See Machine-to-Machine (M2M)

M2P
See Machine-to-Person (M2P)

MAC/Macintosh
Personal computer manufactured by Apple

MAC Address
Media Access Control (MAC) addresses are unique device identifiers stored in the device's read-only memory ... can be used as a unique network address

Machine Capacity
Number of pieces or unites that can be produced by a machine in a given timeframe

Machine Data
Digital data and information generated by a device or piece of equipment

Machine Interface
Boundary between a machine and another machine, device, or human ... See Human Machine Interface (HMI)

Machine Time
From start to finish the time required for a machine to produce a product without human hands-on intervention ... See Manual Time and Machine Work

THE ABIDIAN IIOT/CI DICTIONARY

Machine Work
Work that is done by a machine or piece of equipment as opposed to manual work performed by a human ... See Manual Work

Machine-to-Machine (M2M)
Communications between one machine or piece of equipment and another

Machine-to-Person (M2P)
Communications between one machine or piece of equipment and a human

Main Effect
The change in a response that occurs when a factor is changed from its low level to its high level

Maintainability
Ease and speed with which a maintenance activity can be accomplished ... may be quantified by Mean Time To Repair (MTTR)

Maintenance
Via a value-added partnership providing the highest equipment availability at the lowest cost (highest profitability) to safely operate equipment to meet customer and quality needs ... helps to ensure availability (as opposed to performing repairs)

Maintenance Excellence
Maintenance pacesetter characterized by online predictive monitoring, 98+% uptime, 95+ of all work planned and scheduled, customers setting priorities one week in advance, and costs declining great than 10% per year

Make-to-Order (MTO)
Finished products that are produced because of a specific customer order

Make-to-Stock (MTS)
Finished products that are produced and then stored in inventory

Malware
Malicious software application that is designed to damage a computer or computer system

Manage by Objectives (MBO)
Leadership sets targets, often annually, for the organization to achieve

Management by Wandering Around (MBWA)
Term popularized by Tom Peters and Robert Waterman ... essentially, go look and see ... See *Gemba* and *Genchi Genbutsu*

Manual Time
Hands-on time required for an operator to complete a given task ... See Machine Time and Manual Work

THE ABIDIAN IIOT/CI DICTIONARY

Manual Work
Work that is performed by a human as opposed to work performed by a machine ... See Machine Work and Manual Time

Manufacturing Cycle Efficiency (MCE)
Ratio of value added time to lead time ... See Lead Time and Value Added Time (VAT)

Manufacturing Technology
Equipment, knowledge and systems used to produce products or other equipment

Mass Production
Batch operation approach to production in which large lots (batches) of items are processed and moved to the next process regardless of when or whether they are required ... often seen as the production of a large number of identical parts or products ... objects typically wait in a queue until they can be further processed ... commonly referred to as Traditional Manufacturing ... See Assembly Line and Batch Operation

Master Black Belt (MBB)
Six Sigma, Lean Six Sigma, or quality and continuous improvement professional charged with strategic corporate/enterprise-wide process improvement ... the Master Black Belt is qualified to teach and mentor others and is considered a process improvement technical resource to an organization ... See Black Belt (BB)

Material and Information Flow Diagram (MIFO)
Graphical representation of the actions steps (value stream) involved in the material and information flows used to bring a product or service from order entry to customer delivery ... early predecessor to modern value stream mapping ... See Value Stream Mapping (VSM)

Material Flow
Movement of raw materials, work in process, and finished goods through a value stream ... See Value Stream Mapping (VSM)

Material Requirements Planning (MRP)
Software system to manage production planning, scheduling, and inventory ... See Enterprise Resource Planning (ERP)

Matrix Diagram
Graphically organizes large quantities of characteristics, functions, and/or tasks to allow for easier comparison ... can also depict the strength and direction of interrelationships

MBB
See Master Black Belt (MBB)

MBO
See Manage by Objectives (MBO)

MBWA
See Management by Wandering Around

MCE
See Manufacturing Cycle Efficiency (MCE)

MCU
See Microcontroller Unit (MCU)

Mean
Sum of all values divided by the total number of values ... mean for a population is typically expressed with the Greek letter μ ... a sample mean is typically expressed by $\bar{x}$ (pronounced "x bar")

Mean Square
Variance found by dividing the sum of the squares of a variable by the corresponding degrees of freedom ... used in analysis of variance ... See Analysis of Variance (ANVOA)

Mean Time Between Failures (MTBF)
Elapsed time between failures ... calculated as the arithmetic mean (average) between system failures ... sometimes stated as the average time between failures after they are repaired and returned to service

Mean Time To Repair (MTTR)
Equal to the total equipment downtime in a given period divided by the number of failures in that period ... some practitioners include start-up, turnover, and run-in time for completeness

Measure
In Six Sigma, gaining a detailed understanding of a process current state through the collection of reliable data ... 2nd step in the Six Sigma DMAIC process ... See Six Sigma (6σ)

Measurement Scale
Indicates how variables are categorized, counted, and measured ... the four types of scale are interval, nominal, ordinal, and ratio

Measurement System Analysis (MSA)
Study of a measurement system to determine its reliability, repeatability, and accuracy

Mechatronics
Combining computer, controls, electrical, mechanical, telecommunications, and systems engineering ... combination of the words "mechanical" and "electronic"

Median
Midpoint of a data array

Mesh
Ad-hoc local network where individual nodes communicate directly with each other … signals are said to bounce out from the center node or center transmitter from node to node to the outer-most nodes

Message Oriented Middleware (MOM)
Prioritizes the sending and receiving of messages between two or more systems

Message Queuing Telemetry Transport (MQTT)
Industrial Internet of Things (IIoT) machine-to-machine communications protocol … See Machine-to-Machine (M2M)

Metric
In continuous improvement, a carefully monitored key performance indicator linked to bottom-line goals and objectives … See Key Performance Indicator (KPI)

Microcontroller Unit (MCU)
Computer (including central processing unit, memory, and basic input/output) on a single chip

Midrange
Sum of the lowest and highest data values divided by two

MIFO
See Material and Information Flow Diagram (MIFO)

Milk Run
Making frequent pick-ups and drop-offs at many stations and facilities … collect and deliver supplies, products, and materials at fixed times through the use of kanbans … see also Kanban

MIMO
Multiple-Input and Multiple-Output (MIMO)

Min/Max Inventory
Inventory replenishment strategy in which a minimum level of inventory becomes the reorder point and the maximum level the order ceiling … orders are placed to maintain inventory between the minimum and maximum levels

Mind Map
Graphic used to visualize and organize information … central theme is typically in the middle with supporting branches and themes drawn as roots from the center

THE ABIDIAN IIOT/CI DICTIONARY

Minomi
Japanese term (みのみ pronounced me-no-me) literally translated as main part ... technique to deliver raw materials and supplies without any packaging or containers

Minor Stop
During a Minor Stop, 1) flow of the product stops, 2) the operator is required to reset the work piece and make adjustments, and 3) the operator must reactivate or restart the process ... some also consider any operator intervention because of slow line speeds or quality defects to be a minor stop

Mirai
Japanese term (未来 pronounced me-rah-ee) meaning "the future" ... malware targeting Internet of Things (IoT) devices ... See Botnet and Malware

Mistake Proofing
See *Poka Yoke*

Mobile Personal Emergency Response System (mPERS)
Wearable that enables the user to contact help in an emergency ... See Personal Emergency Response System (PERS)

Modal Class
Statistical class with the largest frequency

Modbus
Programmable Logic Controller (PLC) communications protocol that has been adapted to connect industrial devices between a master and up to 247 slave devices

Mode
Value occurring most often in a data set

MOM
See Message Oriented Middleware (MOM

Monte Carlo Simulation
Statistical simulation technique using random numbers

Monument
A design, scheduling, or production step with large scale requirements and lengthy changeover times dictating waiting in a queue for processing ... See Bottleneck and Pacemaker

Motion
Excessive walking, double handling, poor ergonomics, and lost motion ... body movement not adding value ... one of the 8 wastes ... See 8 Wastes

Motor
Electric device that converts electrical energy into mechanical energy

mPERS
See Mobile Personal Emergency Response System (mPERS)

MQTT
See Message Queuing Telemetry Transport (MQTT)

MRP
See Material Resource Planning (MRP)

MTConnect
Open, royalty-free standard within manufacturing for the one-way (read-only) retrieval of process information from numerically controlled shop floor machine tools

MTO
See Make-to-Order (MTO)

MTS
See Make-to-Stock (MTS)

Muda
Japanese term (無駄 pronounced muu-dah) for waste: an activity consuming resources without creating value for the customer ... some break down Type 1 *Muda* as non-value added, but necessary (i.e., Business Value Added) and Type 2 *Muda* as both non-value added and unnecessary ... See 3 Ms, Business Value Added (BVA), and 8 Wastes

Multi-Machine Handling
When an operator is responsible for operating more than one machine

Multi-Skilled
An employee capable of operating multiple processes and pieces of equipment

Multinomial Distribution
Probability distribution for an experiment in which each trial has more than two outcomes

Multiple-Input and Multiple-Output (MIMO)
Radio utilizing multiple antenna receivers and multiple antenna transmitters

Multi-Voting
Technique based on team member desires for quickly narrowing options and setting priorities ... team members cast multiple votes to identify the most popular options and solutions

Mura
Japanese term (斑 pronounced muu-rah) for unevenness and inconsistency ... an uneven pace causing workers to first wait and then hurry ... See 3 Ms

Muri
Japanese term (無理 pronounced muu-ree) for overburdening equipment or workers by requiring them to operate at a higher or harder pace with more force and effort for a longer period of time than equipment design and realistic management practices might suggest ... See 3 Ms

Mutually Exclusive Events
Probability events that cannot occur simultaneously

N

Nagara Switch
From the Japanese term *Nagara* (生産平滑化 pronounced nah-gah-rah) meaning to smooth the flow ... specially designed switch that can be operated by the quick swipe of an operator's hand ... allows an operator to quickly activate a piece of equipment as she moves on to the next piece of equipment

Nagara System
From the Japanese term Nagara (生産平滑化 pronounced nah-gah-rah) meaning to smooth the flow ... system in which seemingly unrelated tasks are operated by the same operator

Natural Team
Team composed of members from a single work unit or process who typically work together

Nearable
Tracking device or beacon that allows a smart device to interact with objects in its close vicinity

Near Field Communications (NFC)
Short-range wireless communications between devices (e.g., paying for a retail item with a smartphone)

Negative Relationship
As one variable increases, a second corresponding variable decreases, and vice versa

Negawatt
Amount of energy saved (rather than consumed) ... used to promote environmental awareness

Nemawashi
Japanese term (根回し pronounced nay-mah-wah-she) for the process of gaining approval ... building consensus (often behind the scenes)

Net Present Value (NPV)
Tool for evaluating current economic value from a series of expenditures and revenues over time

Neural Network (NN)
Computer program capable of learning and adapting ... See Artificial Neural Network (ANN)

New United Motor Manufacturing, Inc. (NUMMI)
Now dissolved joint venture between General Motors and Toyota

NFC
See Near Field Communications (NFC)

NN
See Neural Network (NN)

Node
Point of connection or intersection in a network

Noise
Inherent variability in a process ... represents the change in a response when no change in the factor is made

Non-Conformance
Product or service that deviates from customer and internal quality requirements ... also known as a defective or a discrepancy item ... See Defects

Non-Destructive Testing (NDT)
Testing or inspection monitoring in which the equipment or process is not harmed or damaged

Non-Value Added Activities
Resources like support labor, office spaced, shop space, and offline spares used in a process that the customer has no need for ... activities not directly satisfying the end customer ... activities the end customer does not want to pay for ... See Value Added Activities and Business Value Added

Nonparametric Statistics
Branch of statistics for non-normal distributions

Normal Curve
See Normal Distribution

Normal Distribution
Continuous, symmetric, bell-shaped distribution of a variable ... sometimes referred to as the Normal Curve ... the Standard Normal Distribution is a special normal distribution in that it has a mean equal to zero (0) and a standard deviation equal to one (1) ... see Standard Normal Distribution

Null Hypothesis
Statistical hypothesis stating there is no difference between a parameter and a specific value or that there is no difference between two parameters

NUMMI
See New United Motor Manufacturing, Inc. (NUMMI)

NVA
See Non-Value Added Activities

THE ABIDIAN IIOT/CI DICTIONARY

O

OAE
See Overall Asset Efficiency (OAE)

Obeya
Japanese term (大部屋 pronounced oh-bay-yah) for great room ... a large, highly visual room (or office area) that promotes communication, teamwork, and faster decision making ... sometimes known as the Open Room Effect

OCE
See Overall Classroom Effectiveness

OEE
See Overall Equipment Effectiveness

Ogive
Graph representing the cumulative frequencies for the classes in a frequency distribution

Ohm
Unit of electrical resistance

Ohm's Law
Equation expressing the relationship between electrical voltage, current, and resistance where voltage V is equal to the current I multiplied by the resistance R (i.e., V=IR)

OMCD
See Operations Management Consulting Division (OMCD)

Ohno, Taiichi **(1912-1990)** (大野耐一 pronounced tie-ee-chee oh-no)
Recognized Father of the Toyota Production System and original developer of concepts like the 7 wastes … author of *Toyota Production System: Beyond Large-Scale Production* … See 7 Wastes and Toyota Production System (TPS)

OLAP
See Online Analytical Processing (OLAP)

One Piece Flow
Making and moving one piece at a time with no work in process inventory … an ideal more often seen in books and classrooms than in practice; nonetheless, a worthy goal to strive for … See 3 Principles of Lean and Work in Process (WIP)

One Point Lesson (OPL)
Short, less than 5-minute study lesson or visual covering a particular aspect of a problem or solution … may also be known as a Single Point Lesson (SPL) … many practitioners use to document standard work in an expedient and concise way … See Standard Work

One-Tailed Test
Test indicating the null hypothesis should be rejected when the test statistic value is in the critical region on one side of the mean

One Touch Exchange of Die (OTED)
Reduces tooling exchange and subsequent machine setup and changeover through consolidation of multiple steps/parts into a single operator task … See SMED and Smart Changeover

One-Way ANOVA
Test for differences among means for a single independent variable when there are three or more groups … See Analysis of Variance (ANOVA)

Online Analytical Processing (OLAP)
Web-based software used to analyze data relationships

Online Checks
Four-step method [1) Define Major Components, 2) Identify Historic Failures, 3) Identify Ways to Monitor, and 4) Ensure Completeness] to develop real-time operating, safety, environmental, quality, throughput and maintenance checks … See Autonomous Maintenance and Condition Based Monitoring

OODA
Mnemonic for Overserve Orient, Decide and Act

Open Source
Decentralization of effort that encourages open peer collaboration … resulting product, including source code, is usually freely available to the public

Open-Ended Distribution
Frequency distribution with no specific beginning value or specific ending value

Operating Excellence
Business philosophy or culture dedicated to continuous improvement through the reduction of waste, lead times, and the cash-to-cash cycle

Operating System (OS)
Software that controls and manages a computing device's most basic functions

Operational Definitions
Clear and precise work instructions … See Standard Work

Operations Management Consulting Division (OMCD)
With a goal of ensuring global consistency, an internal Toyota division responsible for spreading the Toyota Production System (TPS) … See Toyota Production System (TPS)

OPL
See One Point Lesson

Ordinal Data
Data that can be ranked despite no precise differences between the ranks (e.g., pretty, prettier, and prettiest)

OS
See Operating System (OS)

OSE
See Overall Service Efficiency

OTED
See One Touch Exchange of Die (OTED)

Outlier
Extreme value in a statistical data set

Output
Signal or data transmitted from or by a device

Output Y's
See Y Data/Values

Outside-In Thinking
Customer-centric focus rather than internal focus on growth and effectiveness … See Customer-Centric

Over Processing
Excessive paperwork, unnecessary approvals, poor process designs, and other productivity impediments leading employees to work harder than necessary to meet customer requirements ... one of the 8 wastes ... See 8 Wastes and 7 Wastes

Over Production
Producing more, sooner or faster than is required by the next process or customer ... considered to be the worst of the 8 Wastes because it has additional costs (e.g., added storage, handling, and possibly, disposal) of its own and generates, hides, or compounds the other seven wastes ... See 7 Wastes and 8 Wastes

Overall Asset Efficiency (OAE)
Same as Overall Equipment Effectiveness (OEE) except the word "Asset" is substituted for Equipment ... See Overall Equipment Effectiveness (OEE)

Overall Classroom Effectiveness (OCE)
Academic classroom productivity metric ... similar to Overall Equipment Effectiveness (OEE) in that it is the product of performance (i.e., speed), availability, and quality (e.g., successes vs. failures) ... See Overall Equipment Effectiveness (OEE)

Overall Equipment Effectiveness (OEE)
Measure of how effectively equipment is being utilized ... obtained by multiplying the elements of performance, availability, and yield ... See Budgeted Production Efficiency (BPE) and Real-time Production Efficiency (RPE)

Overall Service Effectiveness (OSE)
Administrative or service process productivity metric ... similar to Overall Equipment Effectiveness (OEE) in that it is the product of performance (i.e., speed), availability, and quality (e.g., returns and revisits) ... some use Overall People Efficiency (OPE) or Overall Labor Efficiency (OLE) ... See Overall Equipment Effectiveness (OEE)

P

p-Chart
Control Chart used to analyze the proportion of items in a sample that are defective … assists in determining whether a process is in or out of control

P-F Interval
Time between a potential failure and a functional failure … used in setting preventive, predictive, and condition-based monitoring schedules

P-Value
Actual probability of getting the sample mean value if the null hypothesis is true

Pacemaker
Establishes the production pace or speed of a product or service … ideally, should approximate Takt Time … See Takt Time (T_T) and Monument

Pack Out Rate
See Pitch

PAN
See Personal Area Network (PAN)

Packet Switching
Digital network communications technique that groups transmitted data into blocks (known as packets)

Parallel
Components are connected to the same two points in an electrical circuit

Parametric Tests
Statistical tests for population parameters such as means, variances, and proportions involving assumptions about populations from which the samples were selected

Pareto Analysis (Pareto Chart)
Graphical tool (Bar Chart) that helps break a big problem down into parts and then identifies which parts are the most important/frequent ... uses horizontal/vertical bars to represent frequencies

Pareto Principle
Sometimes referred to as the 80/20 rule (e.g., 80% of sales come from 20% of the customers and 80% of classroom disruptions come from 20% of the students)

Pareto, Vilfredo (1848-1923)
Italian scientist and philosopher best known for his development of the Pareto Principle (80/20 rule)

Parkinson's Law
Largely unscientific belief that the available work will expand so that it consumes the available time

PCE
See Process Cycle Efficiency (PCE)

PD
See Program Development (PD)

PDCA
Plan, Do, Check, and Act ... short, iterative process for planning, executing, evaluating, and implementing improvement ... sometimes written as Plan, Do, Study, and Act (i.e., PDSA) ... originally known as the Shewhart Cycle, but is now more commonly known as the Deming Cycle

PDCA Problem Solving
Problem solving method in which an individual or team defines a problem, identifies the root cause, brainstorms improvements, and validates the improvements ... See PDCA

PDF
See Probability Density Function (PDF)

PdM
See Predictive Maintenance (PdM)

PDPC
See Process Decision Program Chart

PDSA
Plan, Do, Study, and Act ... variation of PDCA ... See PDCA

Pearson Correlation Coefficient
Characterizes the strength of a relationship between normally distributed variables

Performance
Comparison of a piece of equipment's or process' speed or actual output with historical best-demonstrated production during an equivalent timeframe

Peripheral
Computing device that operates independent of a computer or CPU but is connected to it (e.g., a printer)

Permutation
Arrangement of n objects in a specific order

PEC
See Process Evaluation Checklist (PEC)

PERS
See Personal Emergency Response System (PERS)

Personal Area Network (PAN)
Collection of connected devices operating in the range of one person (i.e., about 10 meters)

Personal Emergency Response System (PERS)
Device or system that enables a person to contact help in an emergency ... also known as Personal Safety Wearable ... See Mobile Personal Emergency Response System (mPERS)

PERT Chart
See Project Evaluation and Review Technique Chart

Peter Principle
In theory, candidates for promotion are evaluated on their performance in their current job assignment (rather than on performance attributes desirable in the new assignment) ... hence, employees (and managers in particular) "rise to their level of incompetence"

PFEP
See Plan for Every Part (PFEP)

Phi Correlation Coefficient
Characterizes the strength of a relationship between truly dichotomous variables

Photocell
Device that converts light to a voltage or current

Physical Web
Open standard, originally from Google, that allows Internet of Things (IoT) devices to readily and seamlessly interact via web addresses ... intent is to reduce the need for dedicated device apps

Pictograph
Graph using pictures or symbols to represent data

Piece Price
Total cost of a single production unit

Pillar
A major area of focus or emphasis used in tools such as Total Productive Maintenance and Total Productive Manufacturing ... there are usually about 8 pillars ... See Total Productive Maintenance and Total Productive Manufacturing (TPM)

Pin-Pan-Pon
An auditory alarm or signal (i.e., an *andon*) to quickly provide the status of a device or process ... See *Andon*

Pitch
Amount of time required in a production area to make one container of product ... calculated by multiplying the Takt Time (T_T) by the container quantity ... sometimes referred to as the Pack Out Rate

Plan for Every Part (PFEP)
Information detailing the specific details, characteristics, and attributes of a specific part or product

Planning
Process of evaluating the actions and resources required to safely and efficiently perform a given task ... See Scheduling

Platform
Parent or major piece of technology or software (e.g., and operating system) upon which subordinate or smaller applications are developed ... See Operating System (OS)

PLC
See Programmable Logic Controller (PLC)

Plug-in
Electrical: device that is capable of being plugged into an electric wall outlet, circuit, or circuit board ... Computing: accessory software application that extends the capabilities of a parent software application

PM
See Preventive Maintenance

PMTS
See Predetermined Motion Time System (PMTS)

PoE
See Power over Ethernet (PoE)

Point *Kaizen*
Improvement event activity focused solely on a single workstation, task, or operation ... See *Kaizen* and *Kaizen* Event

Point of Use
Material, supplies, and work in process are staged or delivered to the (precise) location where they are consumed or used

Poisson Distribution
Probability distribution used when n is large and p is small and the independent variables occur over time

Poka Yoke
Japanese term (ポカ 避け pronounced poe-kah yoe-kay) for mistake-proofing ... method of mistake or error proofing equipment and processes to catch or prevent inadvertent errors and omissions before they result in defects

Policy Deployment
See *Hoshin* and *Hoshin Kanri*

Population
Totality of all statistical subjects possessing certain common characteristics

Population Correlation Coefficient
Value of the correlation coefficient computed by using all possible pairs of data values (x,y) taken from a population

Positive Relationship
Relationship between two variables such that as one variable increases, the other increases as well (and vice versa)

Positively Skewed Distribution
Majority of data values fall to the left of the mean

Power Cycle
Turn off, wait a few seconds, and then restore power

Power over Ethernet (PoE)
Uses Ethernet cable to deliver low-voltage power to a device (e.g., passive IP cameras)

Power over Wi-Fi (PoWi-Fi)
Converts signal transmitted by a Wi-Fi router into direct current

Power Supply
Source of electric energy such as a battery, USB port, or electric wall outlet

PPAP
See Production Part Approval Process (PPAP)

PPM
Parts per Million ... See Defects Per Million Opportunities (DPMO)

Predetermined Motion Time System (PMTS)
Technique to break human motion into short, elementary movements ... See Therblig Analysis

Predictive Maintenance (PdM)
Measuring or monitoring the condition of a piece of equipment or process to assess its condition and likelihood of failure during some future time period and then acting to prevent the failure and/or the consequences ... assessing equipment or process condition in a proactive way

Preventive Maintenance (PM)
Performing time or intensity based tasks such as lubrication to sustain the life of a piece of equipment or process

Principles of Lean
1) Define value from the customer's perspective, 2) Identify the steps in the value stream, 3) Eliminate waste, 4) Flow the process, 5) Pull the product, 6) Involve and empower employees, and 7) Pursue to perfection ... See Lean and 3 Principles of Lean

Prioritization Matrix
Prioritizes tasks, issues, characteristics, or possible actions based on known, weighted criteria ... matrix axes are usually benefit and cost

Proactive Maintenance
Initiatives taken to improve equipment reliability such as a dedicated preventive/predictive maintenance team, written standard work practices, defined training programs, computerization, work planning and scheduling, and emergency work less than 5% of all work

Probability
Chance of an event occurring

Probability Density Function (PDF)
Distribution of relative frequencies representing the probability of obtaining a particular value or outcome ... can be either discrete or continuous

Probability Tree
Used to determine the likelihood of individual scenarios

Problem Solving
Process of defining a problem, identifying solutions, selecting the best solution, and implementing the same ... within Lean, a problem is often defined as a goal or objective that has not yet been realized ... solutions should be win-win for customers, company, and employees ... See A3 Problem Solving and Root Cause Failure Analysis (RCFA)

Process
Series of production activities, material, and information flow that transforms inputs into defined product and service outputs

Process Capacity
Maximum quantity of product or services that can be produced by a process during a given amount of time

Process Cycle Efficiency (PCE)
Ratio of value added time and lead time ... See Lead Time and Value Added Time (VAT)

Process Decision Program Chart (PDPC)
Maps events and contingencies ... useful in moving from problem statement to possible solutions in the development of an implementation plan

Process Evaluation Checklist (PEC)
Helps ensure the right order, tools, supplies, materials, people, and associated quantities are in place

Process Map
Block diagram (i.e., flowchart) illustrating the flow of work and the boundaries of a process, its major inputs, and its major outputs ... often used as a mapping tool for Six Sigma (6σ)

Process Mapping
Act of generating a Process Map ... See Process Map

Process Owner
Individual with direct responsibility for process performance and resources ... typically, a team leader, supervisor, or manager

Process Razing
Gaining an understanding of current process conditions across a part family or value stream and then developing ways to make the process steps more common and similar

Product Family
Products or services requiring similar processing steps

Product Family Matrix
A mixed product model for drawing and analyzing value streams incorporating many products ... See Value Stream Mapping (VSM)

Production Part Approval Process (PPAP)
Approval of a new production part before it can be used in manufacturing ... intent is to ensure quality and performance

Production Preparation Process (3P)
Design methodology of looking first at process flow to ensure quality, then what tooling will accommodate the desired flow, quality, and ease of operation, and finally, what is the right-sized equipment to accommodate the tooling and desired results

Production Smoothing
Production scheduling that, over time, takes out fluctuations associated with varying customer orders ... See *Heijunka*

Program Development (PD)
Training Within Industry workshop for trainers and facilitators ... See Training Within Industry (TWI)

Programmable Logic Controller (PLC)
Ruggedized, high-reliability industrial computer adapted for the control of (industrial) processes

Project Evaluation and Review Technique (PERT) Chart
Scheduling and planning tool depicting in a flow chart format project activity interdependencies ... sometimes known as an Activity Network Diagram ... See Activity Network Diagram and Gantt Chart

Protocol
Computer language or syntax used for devices to communicate

Pull System/Production
Production control in which downstream activities signal their needs to upstream activities ... production system activated by customer demand and based on communication of actual, real-time needs ... See Principles of Lean and 3 Principles of Lean

Pulse Width Modulation (PWM)
Using a digital system to approximate an analog signal

Push System/Production
Manufacturing large batches at maximum production rates based on forecast or budget independent of customer demand ... See Mass Production and Batch Operation

Q

QAS
See Quality at the Source (QAS)

QCD
Quality, Cost, and Delivery … See 3 Elements of Demand

QFD
See Quality Function Deployment

QMS
See Quality Management System (QMS)

QRM
See Quick Response Manufacturing (QRM)

QS-9000
Inactive quality standard originally developed by Ford, General Motors, and Chrysler

Qualitative Variable
Variable planned into distinct categories based on some characteristic or attribute

Quality
Meeting all the expectations, specifications, and requirements, stated and unstated, of the customer … See Yield

Quality at the Source (QAS)
Building quality into a process through the implementation of techniques to reduce errors (e.g., mistake proofing) … See *Poka Yoke*

Quality Circle
Typically, a voluntary team of co-workers under the leadership of an elected team leader ... team routinely meets, often self-managed and during paid work hours, to identify and resolve work-related opportunities ... also known as a Quality Control Circle ... See *Ishikawa, Kaoru* (1915-1989)

Quality Control
Collection and analysis of data obtained from small samples of products or services to help identify defects in the process

Quality Function Deployment (QFD)
System for translating customer requirements into appropriate requirements from research, product development and engineering to manufacturing, marketing, sales, and distribution ... See Voice of the Customer (VOC)

Quality Management System (QMS)
Collection of practices and processes designed to consistently meet customer requirements (quality and otherwise)

Quality, Cost, and Delivery (QCD)
See 3 Elements of Demand

Quantified Community
Utilizes sensors to track air quality, energy consumption, pedestrian and vehicular traffic, waste flow, etc.

Quantified Self
Sensors that monitor a person's diet, habits, moods, location, and motion

Quantitative Variable
Numerical variable that can be ordered or ranked

Quartile
Location measure of a data value ... divides a distribution into four groups

Queue Time
Accumulated time a customer, part, or product sits waiting to be processed ... it is not unusual to see queue times on the order of 80% or more of lead time ... See Lead Time and Waiting

Quick Changeover
See Smart Changeover

Quick Response Manufacturing (QRM)
Ability to quickly and effectively respond to changing market demand and product requirements

R

R Chart
Control chart used to analyze the ranges of measurement variables to determine if a process is in or out of control

Radio Frequency (RF)
Electromagnetic "radio" wave frequencies in the range from about 3 kHz to 300 GHz

Radio Frequency Identification (RFID)
Short-range radio wave communications technology involving a small tag and nearby readers

Random Sample
Sample obtained by random or chance methods ... sample for which every member of the population has an equal chance of being selected

Random Variable
Variable whose value is determined by chance ... also known as a random number

Range
Highest data value minus the lowest data value; i.e., difference between the largest and smallest values in a data set

Ransomware
Malware that holds a victim's computing device or data hostage until the victim pays to have it released

Raspberry Pi
Small single-board computer or microcontroller

RCA
Root Cause Analysis ... See Root Cause Failure Analysis (RCFA)

RCFA
See Root Cause Failure Analysis (RCFA)

RCM
See Reliability Centered Maintenance

Reactive Maintenance
Maintenance work characterized by a predominantly repair workload, staffing for emergencies, no or limited preventive maintenance outside of lubrication, and limited performance stewardship ... also known as "firefighting" and Responsive Maintenance ... See Run-to-Failure

Real-time Production Efficiency (RPE)
Modification of the long-term Overall Equipment Effectiveness (OEE) calculation to promote monitoring of real-time production performance using available performance, availability, and yield information

Red Tag
5S label to indicate an item is unneeded and is to be removed from a work area ... usually, item is moved to a holding area for some agreed amount of time before it is disposed of, donated, or sold

Regression
Statistical technique to quantify the relationship between one response variable and one or more predictor variables ... relationship may be positive or negative, linear, or non-linear

Reliability
Dependability or capability of a piece of equipment or process to perform its intended function when called upon

Reliability Centered Maintenance (RCM)
Structured approach to maintenance developed by the military and airline industry to determine equipment maintenance strategies required to ensure a piece of equipment or process continues to fulfill its intended function

Remote Machine Interface (RMI)
Human machine interface and interfaced device are located in different places ... See Human Machine Interface (HMI) and Machine Interface

Repair
Activity returning a failed piece of equipment or process to its original (or sometimes "acceptable") functionality

Repeatability
Portion of the Gage R&R accounting for variation induced by the gage

REpresentational State Transfer (REST)
Method of communicating between computer systems via the internet

Reproducibility
Portion of the Gage R&R accounting for variations induced by the operator

Resistance
Measure of opposition to the flow of electric current (causing electric energy to be transformed into heat) ... See Ohm's Law

Respect for Humanity/People
Deep respect for people and partners ... a key element and often distinguishing characteristic of the Toyota Production System (TPS) ... See Toyota Production System (TPS)

Responsive Maintenance
See Reactive Maintenance

REST
See REpresentational State Transfer (REST)

Return on Investment (ROI)
Income benefits derived from an investment ... $\% ROI = 100 \; x$ *(Total Benefits-Total Cost)/Total Costs*

Rework
Activities required to correct or repair defects ... See Defects

RF
See Radio Frequency (RF)

RFID
See Radio Frequency Identification (RFID)

Right-Sized Equipment
Matching equipment to specific product, job, rate, and space requirements

Right-Tailed Test
Used on a hypothesis when the critical region is on the right side of the distribution

Risk
Potential for unwanted, negative consequences

Risk Priority Number (RPN)
Product generated by multiplying the resulting risk, impact, and frequency ratings generated by a Failure Mode Effects Analysis (FMEA) ... See Failure Mode and Effects Analysis (FMEA)

RMI
See Remote Machine Interface (RMI)

ROI
See Return on Investment (ROI)

Root Cause
The deep, underlying cause of a problem in a process or system ... See Root Cause Failure Analysis (RCFA)

Root Cause Analysis (RCA)
See Root Cause Failure Analysis (RCFA)

Root Cause Failure Analysis (RCFA)
Problem solving methodology focused on identifying the true underlying causes of problems or events ... based on the belief problems are best solved and reoccurrences prevented by attempting to correct or eliminate root causes, as opposed to merely addressing immediately obvious symptoms ... popular techniques include Barrier Analysis, Bayesian Inference, Cause and Effect (Fishbone) Diagrams, Causal Factor Tree Analysis, Change Analysis, Failure Mode and Effects Analysis (FMEA), 5 Whys, Kepner-Tregoe, and Pareto Analysis

Router
Device that receives and re-transmits (re-routes) data packets

Run Chart
Charting performance over time ... See Control Charts

Run-to-Failure
Strategy involving little to no maintenance in which the equipment or process is operated until it fails ... results in reactive maintenance (typically, the most expensive form of equipment maintainability) ... sometimes referred to as "Firefighting" ... See Reactive Maintenance

SaaS
See Software as a Service (SaaS)

Safety Stock
Work-in-process or finished goods accumulated to prevent downstream customers from being starved by upstream process capability issues ... sometimes referred to as Emergency Stock

Sample
Subgroup or subset of a statistical population

Sample Size
For Continuous Data, the sample size can be calculated as $n = \left(\frac{1.96s}{\Delta}\right)^2$ and for Discrete Data as $n = \left(\frac{1.96s}{\Delta}\right)^2 P(1-P)$ where n is the sample size, s is the standard deviation, Δ is the desired level of precision, 1.96 is a constant representing a 95% confidence level, and P is an estimate of the proportion of the process or population that is defective ... See Sample and Sampling

Sampling
Taking data on one or more subsets of a larger group (population) to characterize the larger group ... best sampling techniques tend to be both random and systematic (every n^{th}) ... See Bias and Sampling Error

Sampling Error
Difference between a sample measure and the corresponding population due to the sample not being a perfect representation of the entire population

Sankey Diagram
Flow diagram, typically software-based, where the width of the connecting lines is proportionate to the flow quantity

SCADA
See Supervisory Control and Data Acquisition (SCADA)

Scatter Plot
Graph of interdependent and dependent variables in regression and correlation analysis ... See Correlation and Regression

Schedule Compliance
Number of scheduled activities (e.g., work or customer orders) completed during a given time period divided by the total number of activities scheduled ... See Scheduling

Scheduling
Act of synchronizing multiple work plans to ensure the involved resources and negative production impacts are both minimized and optimized ... See Planning

Scientific Management
Business management philosophy of using scientific problem solving tools and time/motion studies (rather than operator knowledge) to improve productivity and efficiencies ... also known as Taylorism ... See Taylorism and Taylor, Frederick Winslow (1856-1915)

SCO
See Successful Customer Outcome

Scheffé Test
Should the null hypothesis be rejected, a post-ANOVA test to locate significant differences in the means ... See Analysis of Variance (ANOVA)

Scrap
Finished product and work in process that cannot be reworked or repaired to bring it to customer requirements and specifications

Scrum
Five step framework most often used in software project management

Semiconductor
Device typically made of silicon, germanium, or gallium that conducts electricity in some conditions but not others ... a component of most electronic circuits

Sensei
Japanese term (先生 pronounced sin-say) for learned teacher or master

Sensor
Device used to detect or measure physical characteristics (e.g., luminance, motion, temperature, or humidity) and convert it to voltage, current, or signal

THE ABIDIAN IIOT/CI DICTIONARY

Sequential Changeover
Changeover or transition tasks are performed sequentially; i.e., tasks are performed in order or sequence one after another

Serial Communication
Serial protocol between two devices

Serial Peripheral Interface (SPI)
Short-range communication protocol between sensors and a microcontroller

Series
In an electrical circuit, devices are in series when current flows from one device to the other

Server
Computing device that provides and controls access or functionality to other devices (who are known as clients) ... See Client

Setup Reduction
See Smart Changeover

Shadow Board
Tool storage area, often a peg board, where the location for each tool is identified by an outline (i.e., shadow) of the tool ... See Visual Workplace

Shainin System
Collection of 20 statistical tools used in the quality improvement field

Shewhart Cycle
See PDCA

Shewhart, Walter (1891-1967)
Best known as the Father of Statistical Quality Control (SQC) ... See Statistical Quality Control (SQC)

Shield
Microcontroller expansion board

Shingo, Shigeo **(1909-1990)** (新郷 重夫 pronounced she-ghee-oh sheen-go)
Developed the concepts of *Poka Yoke* (i.e., Mistake Proofing) and Single Minute Exchange of Die (i.e., Smart Changeover) ... in 1988, Utah State University recognized *Shingo* for his lifetime accomplishments and created the Shingo Prize recognizing world-class Lean initiatives

Ship to Line
Incoming raw material and supplies are brought directly from the supplier to the production area ... intent is to reduce transportation and storage costs

Shojinka
Japanese term (少人化 pronounced show-jeen-kah) for flexible workforce or workplace

Short Circuit
Fault between an energy source and ground ... also known as a short

SIGFOX
Low-bandwidth wireless network technology that connects low-energy objects ... devices tend to be on and continuously transmitting small amounts of data (e.g., a smartwatch)

Sigma
Designates the distribution or spread about the mean of a process or product characteristic ... one standard deviation from the mean ... See Mean

Sigma Value
Indicates how often defects are likely to occur ... the higher the Sigma Value, the less likely a process will produce defects ... to a limit, as sigma increases, costs go down, cycle time goes up, and customer satisfaction goes up

Signal-to-Noise Ratio
Comparison of the power (strength) of a desired signal to the power of surrounding background noise

SIM
See Subscriber Identity Module (SIM)

Simulation Techniques
Using probability statistics and experiments to approximate reality

Single Minute Exchange of Die (SMED)
See Smart Changeover

Single Point Lesson (SPL)
See One Point Lesson (OPL)

SIPOC
Supplier-Input-Process-Output-Customer diagram ... useful in capturing and displaying system flow from raw material to finished product

Six Sigma (6σ)
Statistical methodology pioneered at Motorola to reduce defect levels (errors) below 3.4 Defects Per Million Opportunities (DPMO) ... Six Sigma methods can be applied to virtually any manufacturing, administrative, service, or transactional process ... See Black Belt (BB), Master Black Belt (MBB), Six Sigma Performance and Six Sigma Business Philosophy

Six Sigma Business Philosophy
Business approach centered around a largely statistical, fact-based toolset that involves an infrastructure of "belted" improvement professionals driving continuous improvement ... See Six Sigma and Six Sigma Performance

Six Sigma Performance
Operating at or below a defect rate of 3.4 defects per million opportunities ... See Six Sigma (6σ) and Six Sigma Business Philosophy

Six Step Problem Solving
Problem-solving technique of 1) Identify problem, 2) Analyze, 3) Identify multiple solutions, 4) Select the best solution, 5) Implement, and 6) Evaluate and verify

Skills Matrix
Table (i.e., matrix) displaying employees and the status of their mastery of various skills

SKU
Stock Keeping Unit ... a part or product number

Smart
Within computing: a device, machine, or system with some level of self-awareness or intelligence

SMART
Specific, Measurable, Attainable, Relevant and Time-based ... criteria often used as a guide in goal setting

Smart Building
Connected, intelligent facility that is designed to reduce energy costs and environmental impact while increasing productive interactions with its human inhabitants

Smart Changeover
Process of transitioning equipment from one production run to another in as little time and waste as safely possible ... originated in the automotive industry where the goal was to change large mega-ton dies in less than ten minutes (i.e., Single Minute Exchange of Dies (SMED) ... sometimes known as Quick Changeover

Smart Device
Instrument, machine, or piece of equipment with integrated computing capabilities ... also known as an intelligent device ... See Intelligent Device

Smart Grid
Use of intelligent devices (e.g., smart meters) on the electric grid ... See Intelligent Device and Smart Meter

THE ABIDIAN IIOT/CI DICTIONARY

Smart Manufacturing
Interacting with a digital representation or clone of physical devices ... allows managing a machine without being in the immediate physical area of the machine ... initiative encouraging more open connectivity between smart equipment, facilities, products, and processes ... See Industrial Internet of Things (IIoT).

Smart Meter
Intelligent device that measures and transmits flow, temperature, pressure, physical characteristic, or other types of consumption data ... often associated with a utility such as water, gas, or electricity or temperature and density-type physical characteristics in a manufacturing environment ... See Intelligent Device

Smart Watch
Wrist watch with some level of intelligence that can communicate wirelessly and interacts with the wearer

Smartphone
Mobile handheld telecommunications device with an intelligent operating system

SMED
Single Minute Exchange of Die ... See Smart Changeover

Smith, Bill (1929-1993)
Engineer and Motorola quality assurance professional and vice president widely recognized as the Father of Six Sigma

Smoothing
See *Heijunka*, Level Scheduling, and Production Smoothing

SoC
See System on a Chip (SoC)

Software as a Service (SaaS)
Fee-based subscription (rather than upfront purchase) for software applications

Spaced Repetition
Technique to improve memory or brain retention ... See Quantified Self

Spaghetti Chart/Diagram
Diagram depicting the path an employee, material, or product travels along the steps of a value stream ... used to identify motion and transportation wastes ... See Motion and Transportation

SPC
See Statistical Process Control

Special Cause Variation
Non-random causes of variability typically attributed to influences outside of a given process … can be detected by Control Charts … See Common Cause Variation, Stable Process, and Control Charts

SPI
See Serial Peripheral Interface (SPI)

SQC
See Statistical Quality Control (SQC)

SQDC
Safety, Quality, Delivery, and Cost … a variant of the 3 Elements of Demand … See 3 Elements of Demand

SSID
Serial Set IDentifier: name that identifies a wireless network

Stable/Stability
Lack of variation … See Stable Process

Stable Process
A process is statistically stable when it has only common cause variation … See Common Cause Variation and Special Cause Variation

Standard Deviation
Statistical indication of variability or spread calculated by taking the square root of a variance … can also be calculated by taking the square root of the sum of the squared deviations from the mean for each value, divided by n-1 … expressed by the Greek letter σ for a population and "s" for a sample

Standardization
Act of standardizing work processes … See Standard Work

Standard Normal Distribution
Normal distribution in which the mean is equal to the 0 and the standard deviation is equal to 1

Standard Score
Difference between a data value and the mean divided by the corresponding standard deviation

Standard Work
Established by consensus, a structured and repeatable process for (precisely) accomplishing work based on Takt Time, work sequence, and standard inventory and equipment ... a "standard" and not a "guideline" in doing work: standard work deviations in a Lean culture are considered an abnormality and an opportunity for improvement ... See Best Practice, Leader Standard Work (LSW), Standard Work in Process, and Standard Work Practice (SWP)

Standard Work in Process
Amount of work in process (WIP) inventory specified in the process's standard work practices; i.e., the optimum quantity for consistent repetitive operations ... See Standard Work and Work in Process (WIP)

Standard Work Practice (SWP)
Work instruction or procedure established by consensus that documents a structured and repeatable process to ensure safe, consistent results ... See Standard Work

Statistical Process Control (SPC)
Application of statistics to control a process ... often used interchangeably with Statistical Quality Control (SQC)

Statistical Quality Control (SQC)
Application of statistics to control quality in a process ... often used interchangeably with Statistical Process Control (SPC)

Strategic Planning
Process of defining an organization's high-level direction and goals

Stratification Factors
Collection of descriptive information to assist in identifying patterns and root causes

Subscriber
Data end user ... may involve the payment of a fee for service

Subscriber Identity Module (SIM)
Contains the identification, security, and access rules required to authenticate access to a network ... also known as Internal Mobile Subscriber Identity (IMSI)

Successful Customer Outcome (SCO)
Resulting outcome, and the "process" experienced by the customer in obtaining the outcome, that the customer would define as making their life simpler, easier, safer, more profitable or more successful

Suggestion System
Process where workers are encouraged to identify safety, inefficiency, and waste concerns with sometimes individual/team rewards for suggestions resulting in financial savings

Supermarket
Physical location where a preset quantity of material is controlled and released through the use of *kanbans* ... See *Kanban* and Pull System

Supervisory Control and Data Acquisition (SCADA)
Computer-based control architecture for large, geographically-dispersed industrial equipment such as that found in the utility industry or continuous process industries like nuclear, refining, and chemicals

Supplier
Provides material resources or information to a process

Supply Chain
System of companies and activities involved in moving a product or service from one company (the supplier) to another (the customer)

Swim Lane Flowchart
Flowchart emphasizing the "who" in "who does what" ... useful in the analysis of handoffs and transfers

Switch
Device that can operate or close (complete) an electric circuit

SWP
See Standard Work Practice (SWP)

Synchronous Manufacturing
Needed resources and materials to manufacture products to meet customer demand are available just in time ... See Just in Time (JIT)

System of Profound Knowledge
Deming's system theory-based management philosophy consisting of four interrelated components: 1) Appreciation of a system, 2) Theory of knowledge, 3) The psychology of change, and 4) Knowledge about variation ... See Deming, W. Edwards (1900-1993)

System on a Chip (SoC)
Single integrated circuit "chip" that contains all the circuits (i.e., input/output, memory, and central processor) required for a complete computing "system"

THE ABIDIAN IIOT/CI DICTIONARY

T

t Distribution
Family of bell-shaped curves based on degrees of freedom similar to the standard normal distribution with the exception the variance is greater than 1 ... used when testing small samples and when the population standard deviation is unknown

T-Card
T-shaped card, usually color coded on the front and back, for use on a *Kamishibai* Board ... See *Kamishibai* Board and Visual Workplace

t-Test
Statistical test for the mean of a normally distributed population when the population standard deviation is unknown and the sample size is less than 30

Takt Time (T_T)
German term for pulse or beat ... required pace/rhythm: available production time divided by customer demand ... comparing the ever-changing takt time value to production cycle times will help determine if a process is in sync with customer orders ... See Cycle Time (C_T) and 3 Principles of Lean

Takt Time Chart
See Value Added Chart

Task Time Chart
See Value Added Chart

Taylor, Frederick Winslow (1856-1915)
Best known for his development of the theory of Scientific Management ... See Scientific Management

Taylorism
See Scientific Management

TCO
See Total Cost of Ownership (TCO)

TCP/IP
See Transmission Control Protocol/Internet Protocol (TCP/IP)

Team Leader
Employee leading three to five other employees

Tebanare
Japanese term (手離れ pronounced teh-bah-nah-ray) for "hands off" ... goal is to economically automate historically manual machines and their interfaces to allow humans to do work that is suitable for only for people

TEEP
See Total Effective Equipment Performance (TEEP)

Telematics
Long-distance data transmissions technology used to transmit and display information like the position of a vehicle or movable equipment

Tetrachoric Correlation Coefficient
Characterizes the strength of a relationship when both variances have been artificially reduced to two categories

Therblig Analysis
18 standard elements used to analyze operating motion ... See Motion and Gilbreth, Frank (1868-1924) and Predetermined Motion Time System (PMTS)

Thermography
Monitoring and analyzing the condition of equipment and processes through the measurement of heat ... scanning for hot spots, leaks, or insulation breakdown

Theory of Constraints (TOC)
Process using cause and effect thinking to identify and eliminate bottlenecks ... See Bottleneck, Constraint, and Goldratt, Eliyahu (1947-2011)

Theory X
Management theory based on employees are by their very nature lazy, unambitious, and unhappy with their work

Theory Y
Management theory based on employees are motivated, ambitious, enjoy their time at work, and want to do a good job

Theory Z
Japanese management style based on employee and company loyalty, lifetime job security, and the employee's well-being on and off the job

Throughput
Rate of production

Throughput Velocity
Metric that relates total output over a specified time period to work in process (WIP) inventory ... calculated as follows:

$$Throughput\ Velocity = \frac{\frac{\#\ of\ product\ shipped}{Time\ period}}{Total\ WIP\ Invntory}$$

TIM WOODS
Mnemonic to help recall the 8 Wastes: Transportation, Inventory, Motion, Waiting, Over Processing, Over Production, Defects, and Skills/Resources Underutilized ... See 8 Wastes

Time Value Map
Visual depiction of process value added and non-value added time ... See Value Added Time (VAT) and Non-Value Added Time

Total Cost of Ownership (TCO)
Life-cycle cost of a piece of equipment including original purchase, installation, maintenance, and end-of-life disposal

Total Effective Equipment Performance (TEEP)
Another name for 24/7/365 Overall Equipment Effectiveness (OEE)

Total Productive Maintenance (TPM)
Techniques involving largely mechanics and operators to ensure every machine in a production process is able to perform its required tasks ... TPM seeks maximum productivity of equipment and addresses the total life cycle of equipment ... some refer to the techniques as Total Productive Manufacturing in that it requires the total participation of all employees ... See Total Productive Manufacturing (TPM)

Total Productive Manufacturing (TPM)
Recognizing optimization is much more than just the mechanics and operators working close together, some practitioners describe the application of the Total Productive Maintenance (TPM) methodology to the entire manufacturing process as Total Productive Manufacturing ... See Total Productive Maintenance (TPM)

Total Quality Management (TQM)
Business quality management philosophy and toolset pioneered by W. Edwards Deming that focuses on reducing errors, increasing customer satisfaction, streamlining the supply chain, modernizing equipment, and ensuring workers have the highest level of training ... a goal of TQM is to limit errors to 1 per million units produced ... sometimes known as Total Quality Control (TQC) ... See Deming, W. Edwards (1900-1993)

Toyoda, Kiichiro (1894-1952) (豊田 喜一郎 pronounced kee-ee-chee-roe toe-yoe-dah)
Japanese entrepreneur and son of *Sakichi Toyoda* ... his decision to take the family's loom works business into automobile manufacturing led to what would eventually become Toyota Motor Corporation

Toyoda, Sakichi (1867-1930) (豊田 佐吉 pronounced sah-kee-chee toe-yoe-dah)
Son of a Japanese carpenter who became an inventor and industrialist founding Toyota Industries Co., Ltd. ... his most famous invention was the automatic power loom which incorporated the principle of *Jidoka* (i.e., Autonomation) ... See *Jidoka*, Autonomation, and Toyota Production System

Toyota Production System (TPS)
Developed by Toyota Motor Corporation to provide the best quality, lowest cost, and shortest lead time through the elimination of waste ... recognizes the ideal situation is when people and machines work together adding value without creating waste

Toyota Way 2001
Internal document summarizing Toyota's philosophy, values, and manufacturing ideals ... first published in 2001

TPM
See Total Productive Maintenance (TPM) and Total Productive Manufacturing (TPM)

TPS
See Toyota Production System (TPS)

TQM
See Total Quality Management (TQM)

Traceability
Ability to trace the life history of a product from raw material to finished product

Traditional Manufacturing
See Batch Operation and Mass Production

Training Within Industry (TWI)
Initiative originally sponsored by the U. S. Government during World War II to rapidly train those replacing the men in manufacturing who went to fight in the war

THE ABIDIAN IIOT/CI DICTIONARY

Transceiver
Transmitter-receiver that both transmits and receives

Transducer
Changes one form of energy to another ... See Actuator and Sensor

Transmission Control Protocol/Internet Protocol (TCP/IP)
Base end-to-end communications protocol used by the internet ... defines how data should be sorted into packets, addressed, transmitted, routed, and received

Transportation
Unnecessary movement over an extended distance... one of the 8 Wastes ... See 8 Wastes

Traveler Check Sheet
Inspection or status form that travels with a product or service through the value stream

Traveling Turkey
Often humorous tool used in sustaining Standard Work ... highlights when accepted, standard work processes have not been practiced by a team ... See Standard Work

Travel Time
Time it takes an employee to move to another location to pick up or put down parts, tools, supplies, or products ... See Transportation

Tree Diagram
Maps in increasing detail indicate the full range of paths and tasks required to achieve a primary goal and related sub goals ... graphically resembles an organization chart or family tree

Trend
Tendency, over time, for a variable to increase, decrease, or remain unchanged

Tribology
Monitoring the condition of equipment and processes through the analysis of its lubrication and oiling systems

TRIZ
Systematic approach to creating innovative solutions to technical problems

True North
Idiom attributed to Toyota describing the ideal (perfect) state an organization would aspire to

Turnover
See Inventory Turns

T$_T$
See Takt Time (T$_T$)

TWI
See Training Within Industry (TWI)

Two-Bin System
Two bins or containers are used to trigger or signal the re-order and replenishment of parts and materials

Two-Sided Test
Indicates the null hypothesis should be rejected when the test value is in either of the two central regions

Two-Way ANOVA
Tests the effects of two or more interdependent variables and the possible interaction between them

Type I Error
Error that occurs if one rejects the null hypothesis when it is true (i.e., incorrect decision to reject something when it is acceptable)

Type II Error
Error that occurs if one does not reject the null hypothesis when it is false (i.e., incorrect decision to accept something when it is unacceptable)

U

U Chart
Count per unit Control Chart ... See Control Chart

U-Cell
See Cellular Manufacturing

UART
See Universal Asynchronous Receiver Transmitter (UART)

UCL
See Upper Control Limit (UCL)

UEM
See Unified Endpoint Management (UEM)

UL or U/L
See Uplink (UL or U/L) or Upload (UL or U/L)

Unified Endpoint Management (UEM)
Securing and controlling desktop computers, laptops, tablets, and smartphones from a single console

Uniform Resource Identifier (URI)
Unique identifier for internet content

Uniform Resource Locator (URL)
Form of Uniform Resource Identifier allowing an internet browser to access a specific web page ... also known as a web address ... See Uniform Resource Identifier (URI)

Unit
Object on which a measurement or observation can be made

Unilateral Tolerance (One-Sided Specification Limit)
Form of tolerance equal to the minimum or maximum performance specification beyond which a measured performance parameter is considered to have failed

Union of Japanese Scientists and Engineers
Formed by the Japanese government after World War II to foster prompt improvement in Japanese Manufacturing ... today the Union of Japanese Scientists and Engineers administers the international Deming Prize ... See Deming, W. Edwards (1900-1993)

Universal Asynchronous Receiver Transmitter (UART)
Computer interface controller that converts received parallel bytes into a serial stream of bits

Universal Serial Bus (USB)
Computer or electronic device port or connection

Unix
Family of multiuser computer operating systems ... See Operating System (OS)

Unplanned
Activity for which a pre-determined procedure or plan had not been documented, or for which all parts, tools, and labor to carry out the task had not been estimated, or the availability of the latter assured prior to commencement of work ... See Planning

Uplink (UL or U/L)
Process of uploading, transmitting, or sending data ... See Upload (UL or U/L)

Upload (UL or U/L)
Transmit data to a remote system or device ... See Uplink (UL or U/L)

Unscheduled
Work that had not been included in an approved schedule prior to its commencement ... See Scheduling

Upper Control Limit (UCL)
Maximum acceptable value of a mean or range of a sample

Uptime
See Availability

URI
See Uniform Resource Identifier (URI)

URL
See Uniform Resource Locator (URL)

USB
See Universal Serial Bus (USB)

User Experience (UX)
Degree to which a customer (user) is satisfied by an encounter (experience) usually with technology

Utilization
Percentage of total clock or calendar time a machine is actually used ... See Availability

Utility Tree/Matrix
To choose among alternative options by separately evaluating their respective benefits and the probability of achieving those benefits

UX
See User Experience (UX)

THE ABIDIAN IIOT/CI DICTIONARY

V

V2I
See Vehicle-to-Infrastructure (V2I)

V2V
See Vehicle-to-Vehicle (V2V)

Validation
See Verification

Validity
Ability of a feedback instrument to measure the degree to which inferences derived from measurements are meaningful

Value
Worth placed by the customer on a product or service … within Lean, value can only be established by the end customer … See Value Added Activities

Value Added
See Value Added Activities

Value Added Activities
Those activities that directly contribute to satisfying the end customer … some define as 1) customer cares and is willing to pay, 2) physically changes and adds desirable functionality, and 3) done right the first time, every time … See Non-Value Added Activities and Business Value Added (BVA)

Value Added Chart
Visual depiction of process value-added and non-value added time in which cumulative task cycle times for each station and worker are compared to Takt Time … might also be known as a Task Time Chart or Takt Time Chart … See Takt Time (T_T), Non-Value Added Activities, and Value Added Activities (VAT)

Value Added Time (VAT)
Sum of the timed work elements that actually transform a product in a way the customer is willing to pay for ... See Value Added Activities

Value Stream
All the actions, value-added and non-value-added, required to bring a product or service from concept and raw material to customer delivery

Value Stream Management
Eight-step approach of managing change based on Current and Future State Value Stream Maps ... See Current State Map, Future State Map, and Value Stream Mapping (VSM)

Value Stream Manager
Leader responsible for coordinating activities as an organization implements the activities and performance levels identified in their Future State Value Stream Map ... See Future State Map and Value Stream Mapping (VSM)

Value Stream Mapping (VSM)
Graphical depiction of the action steps (i.e., value stream) involved in the material and information flows used to bring a product or service from order entry to customer delivery ... See Current State Map, Future State Map, and Value Stream

Values
Fundamental beliefs driving behavior and decision making

Variable
Characteristic or attribute that can be assigned different values

Variance
Average of the squares of the distribution between each value and the mean; i.e., square of the standard deviation ... change or swing in a process that may alter its expected or desired outcome ... See Standard Deviation

Variation
Change in characteristic, data, or function caused by common causes, special causes, tampering, or regular or systematic changes in output (i.e., structural variation such as the time of the year) ... also known as the dispersion among values in a data set ... See 3 Ms

VAT
See Value Added Time (VAT)

Vehicle-to-Infrastructure (V2I)
Communication between a smart car (vehicle) and surrounding sensors

Vehicle-to-Vehicle (V2V)
Communications between two vehicles or a road station using the 5.9 GHz band

Venn Diagram
Pictorial representation of intersections, common areas, probability concepts, and rules

Venture Capital
Private equity financing, typically in a new or expanding business, in which there is potential growth but higher than normal risk

Verification
Inspecting to determine if products or services conform to standards

Vertical Deployment
Involvement of all levels of an organization

Vibration Analysis
Uses noise, minute movement, imbalances, looseness, and vibration to proactively evaluate equipment condition

Virtual Private Network (VPN)
Extends the functionality, security, data, and applications of a private local network (i.e., intranet) across a public network such as the internet ... See Intranet and Local Area Network (LAN)

Virtual Team
Remotely located individuals affiliated with a common purpose, project, or organization who typically conduct their business via telephone, the internet, or other forms of electronic communication

Visible Light Communication (VLC)
Wireless communications based on visible light (i.e., infrared or near-ultraviolet) rather than radio waves ... See Light Fidelity (Li-Fi)

Vision
Overarching statement clarifying an organization's goals, values, and ideals

Visual Management
See Visual Workplace

Visual Workflow
Application of outside-in thinking, visual workplace and a customer-centric focus to administrative and service areas

Visual Workplace
Use of displays, colors, charting, *andons*, boards, labels, flow lines, and other similar visuals and controls enabling employees to immediately recognize "at a glance" a standard and any deviation from it ... sometimes known as Visual Management ... See Andon and Abnormality Management

Vital Few, Useful Many
Term popularized by Joseph M. Juran to describe the Pareto Principle (80/20 Rule) ... See Juran, Joseph (1904-2008)

VOC
See Voice of the Customer

Voice of the Customer (VOC)
Structured approach to cooperatively learn and prioritize the requirements, expectations, and desires of the customer and then establish how they will be successfully satisfied

Voltage
Difference in electrical potential between two points ... See Ohm's Law

VPN
See Virtual Private Network (VPN)

VSM
See Value Stream Mapping (VSM)

Waiting
Delays leading to idle employees or equipment ... inefficient use of time ... one of the 8 wastes ... See 8 Wastes

WAN
See Wide Area Network (WAN)

Waste
Any activity consuming resources, but creating no value in the eyes of the end customer ... See 8 Wastes

Waste Walk
Walk through a process focusing on identifying and ultimately eliminating inefficiencies and waste ... See Waste

Water Spider
As opposed to a conventional material handler, a water spider makes rounds supplying needed parts and materials, assisting with changeovers and transitions, providing tools, and any other help that might be needed

Watson
Cognitive computer system capable of interacting with and answering spoken questions from humans

Wearable
Computer intelligence integrated into clothing or an accessory that is worn by a human

Weave
Communications layer or protocol allowing Internet of Things (IoT) devices to communicate with each other, the cloud, and smartphones

Web
Slang for the internet

Web Editor
Software used to create and modify website pages

Weibull Curve
See Weibull Distribution

Weibull Distribution
Probability distribution associated with random equipment and process failures … plotting failures versus time often results in a "bathtub curve" with infant mortality failures on the left and wear out failures on the right … See Bathtub Curve

Weighted Ranking
Evaluate competing proposals against the same relative criteria

Weighted Voting
See Weighted Ranking

Weightless
Series of open wireless standards

Whisker
From a Box Plot, displays the minimum and maximum observations within 1.5 IQR (spanning 75th to 25th percentile) … See Box Plot

White Belt
Person who has received an introductory overview to Six Sigma or Lean Six Sigma … training is typically one to two hours … See Yellow Belt, Green Belt (GB), Black Belt (BB) and Master Black Belt (MBB)

Wi-Fi
Wireless Fidelity … wireless technology found in computer networks allowing intelligent devices to communicate

Wi-Fi Protected Access (WPI)
Protocol used to secure wireless data

Wide Area Network (WAN)
Computing or telecommunications network that extends over a large geographic area by using leased telecommunications circuits … a wide area network can connect multiple local area networks … See Local Area Network (LAN)

Wilcoxon-Mann-Whitney Test
Tests the null hypothesis that two populations have identical distribution functions against the alternative that the two differ only with respect to their median

Windows
Personal computer operating system ... See Operating System (OS)

WIP
See Work in Process (WIP)

Wisdom
In the context of the Toyota Production System, wisdom is having the technical knowledge, first-hand deep understanding, and practical insight to know how to apply an improvement tool in a beneficial, common sense way ... See *Genchi Genbutsu* and Toyota Production System (TPS)

WO
See Work Order (WO)

Womack, James P.
With Daniel T. Jones, authored *The Machine That Changed the World* which coined the term "Lean Production" ... Womack was later the founder and now retired Chairman of MIT's non-profit Lean Enterprise Institute

Work Cell
Self-contained unit including several value-adding operations arranged in process sequence ... typically in a C or U configuration with counterclockwise flow ... See Cell and Cellular Manufacturing

Work in Process (WIP)
Parts and intermediate products between processing stops ... in-process inventory

Work Module
See Work Cell

Workforce
Non-supervisory, non-management employees

Work Order (WO)
Document used by maintenance (and other support or service functions like technical support) personnel to manage maintenance, repair, and project tasks

Work Request
Document used by the maintenance function's customers to request or initiate a maintenance task ... work request becomes a "work order" after appropriate review and approval ... See Work Order (WO)

Work Team
See Natural Team

World Class
Subjective standard indicating international excellence; i.e., best of the best

WPA
See Wi-Fi Protected Access (WPI)

Wristop
Contraction of wrist watch (or band) and desktop

X̄ Chart (x Bar Chart)
Control Chart used to analyze the means of a measured value to determine whether a process is in or out of control … pronounced "x bar" chart … See Bar Graph

X Data/Values
Inputs … leading indicators such as raw material specifications, start or arrival time, composition and cost … Six Sigma techniques are used to predict results (Output Y's) based on the Input X's

X Dock
Cross Dock

XML
See Extensible Markup Language (XML)

XMPP
See Extensible Messaging and Presence Protocol (XMPP)

XMPP-IoT
Open Internet of Things (IoT) standard based on the XMPP messaging protocol … See Extensible Messaging and Presence Protocol (XMPP)

THE ABIDIAN IIOT/CI DICTIONARY

Y

Y Data/Values
Outputs or results ... lagging indicators such as profit, lead time, customer satisfaction, and total defects ... sometimes referred to as Output Y's

***Yamazumi* Chart**
Japanese term (山積み pronounced yah-mah-zuu-me) meaning smoothing or leveling ... a bar chart used to balance operator workload

Yellow Belt
Individual trained in the basic application of Six Sigma/Lean Six Sigma techniques ... they work with Green Belts and Black Belts throughout the project stages and are often the closest to the work ... training is typically several hours to a day or more ... See Green Belt (GB) and Black Belt (BB)

Yield
Comparison of the raw materials put in to a process and the number of products that meet the customer's specifications; i.e., ratio of good products and the total number of units processed

Yocto
Linux development environment used in embedded products ... See Linux

Yokoten
Japanese term (横展開 pronounced yoe-koe-ten) for ensuring all concepts, ideas, and policies are known horizontally across a company

Z

z Distribution
See Standard Normal Distribution

z Shift
Difference between short-term z_{ST} and long-term z_{LT} standard scores ... larger shifts imply a greater ability to control the special factors identified in the subgroups ... See Standard Score

z-Score
See Standard Score

z-Test
Statistical test for means and proportions of a normally distributed population when the population standard deviation is known or the sample size is greater than 30

z Value
Same as z or Standard Score ... See Standard Score

Z-Wave
Automation industry wireless communications protocol (specification) to allow devices in the home (i.e., lighting, access controls, entertainment systems, appliances, etc.) to communicate

Zero Defects
Performance standard and methodology of no defects popularized by Philip M. Crosby ... practically, if commit to watching details and avoiding errors, can approach a goal of zero defects ... see also Total Productive Maintenance (TPM) and Total Productive Manufacturing (TPM)

Zero Quality Control (ZQC)
Mistake proofing approach to manufacturing based on the principle defects are prevented by controlling the process so that it cannot produce defects

Zero UI
Device control via voice or movement without any type of user interface ... See User Interface (UI)

ZigBee
IEEE 802.15.4 wireless protocol allowing home and light industrial automation devices to communicate in small personal area networks with small, low-power digital transmission devices (radios)

ZQC
See Zero Quality Control

Japanese Terms

Many continuous improvement concepts have either originated or were greatly enhanced in Japan. Most of the concepts attributed to Japan have been translated into comparable English terminology, yet the Japanese terms persist. Most the readers of this book will likely be in the English-speaking western hemisphere, and not Japan. Why not use the English version of the word? There are three reasons.

kai   zen

Change   Excellent

Many, sometimes small, common sense changes for the better led by employees in the work area

First, some of the English translations do not do the concept justice. For example, the Japanese word *kaizen* is often simply translated as an "improvement event" versus its literal translation of "excellent change." The two have totally different connotations. "Improvement event" doesn't carry the same enthusiasm and passion for improvement as excellent change. It also doesn't convey the team-oriented, win-win fact-based solutions, focused and engaged employees, nor the long-term success seen with most *kaizens*!

The second reason Japanese terms are used is respect. True continuous improvement is about respect. Deep respect for people is an inherent quality of a successful culture. Success is not in the tools or the technology. Success is in trusting people to implement the tools in practical, common sense ways. Respect and the positive, engaging culture it creates separates many organizations who become successful versus those that merely practice continuous improvement.

The third reason is to avoid error-prone personal biases in the translations of others. Only by going to the original writings to gain a deep personal understanding (*genchi genbutsu* ☺) can we eliminate inadvertent misinformation by others.

135

THE ABIDIAN IIOT/CI DICTIONARY

Out of completeness and respect, this book often uses the English and Japanese terms interchangeably. To distinguish their non-English background, the Japanese terms are denoted with italics.

Some of the more popular Japanese continuous improvement terms and their pronunciation follow on this and the next several pages.[1]

Term	Japanese (*Kanji*)	Pronunciation (Common)	English Translation[2] (Literal)
andon	行灯	ahn-doan	lamp … lantern … light
baka yoke	馬鹿ヨケ	bah-kah yoe-kay	idiot/fool proofing
burabura	ぶらぶら	buu-rah-buu-rah	strolling … wandering
chaku chaku	着々	chah-kuu chah-kuu	load load … steadily
choko choko tomari (or *chokotei*)	長虹堤	choe-koe choe-koe toe-mah-ree (or choe-koe-tay)	sputter, sputter, stop (i.e., a minor stop)
futekisetsu na	不適切な	fuu-tay-kee-say-t'sue nah	inadequate
gemba (or *genba*)	現場	gim-bah (or gane-bah)	site … actual spot … where the action is (where the work is performed)
genbutsu	現物	gane-boot-sue	actual object
genchi	現地	gen-chee	actual place
genchi genbutsu	現地現物	gen-chee gane-boot-sue	actual object … actual place … local spot … genuine/real … go see
genjitsu	現実	gane-jeet-sue	reality … actual (value/price)
gourika	合理化	goe-uu-ree-kah	rationalization, rationalize
hakarigoto	謀	hah-kah-ree-goe-toe	plan … strategy
hanashiai toron	話し合い 討論	hah-nah-she-ee toe-roan	group discussion
hancho	班長	hahn-choe	team leader
hanedashi	はね出し	hahn-dah-she	auto-eject device
hansei	反省	hahn-say	reflection … reconsideration … introspection
heijunka	平準化	high-joon-kah	(production) leveling
hi kōritsuteki	非効率的な	he koe-reet-stay-kee	inefficient
hinshitsu	品質	heen-sheet-sue	material quality

Term	Japanese (*Kanji*)	Pronunciation (Common)	English Translation[2] (Literal)
hitozukuri	人作り	shtoe-zuu-kuu-ree	making people
hoshin kanri	方針管理	hoe-sheen kahn-ree	policy deployment ... strategic planning/management
ichidan	取り調べる	ee-chee-dahn	to investigate ... to examine
iji	維持	ee-jee	maintenance ... preservation
ijou kanri	異常管理	e-joe-uu kahn-ree	abnormality management
Ishikawa, Kaoru	石川馨	kah-oh-ruu ee-she-kah-wah	developer of Quality Circles and the *Ishikawa* (Cause and Effect) Diagram
jidoka	自働化	ji-doe-kah	autonomation with a human touch ... autonomous defect control
jishuken	自習	jee-shuu-ken	autonomy ... self-study or a combination training/improvement event led by management
juugyouin	従業員	juu-g'yoe-uu-een	employee ... worker
kaiaku	改悪	kye-ah-kuu	change for the worse ... deterioration ... a bad change ... opposite of *kaizen*
kaikaku	改革	kye-kah-kuu	radical change/improvement
kaizen	改善	kye-zen	excellent change ... continuous improvement
kikanshi	機関士	kee-kahn-she	engineer
kakushin	革新	kah-kuu-sheen	reform ... innovation
kamishibai	紙芝居	kah-me-she-bye	paper drama or picture show
kanban	看板	kahn-bahn	signal ... sign
kanban hōshiki	かんばん方式	kahn-bahn hoe-she-kee	just-in-time
kangae	考え	kahn-guy	idea, thinking, thoughts, ...
karakuri	絡繰り	kah-rah-kuu-ree	gimmick
karoshi	過労死	kah-roe-she	death from overwork
kata	型	kah-tah	form
katashiki	型式	kah-tah-she-kee	vehicle ... model
keiretsu	系列	kay-ray-t'sue	affiliate ... enterprise
kiken	危険	kee-ken	dangerous
kitanai	汚い	kee-tah-nye	dirty

THE ABIDIAN IIOT/CI DICTIONARY

Term	Japanese (*Kanji*)	Pronunciation (Common)	English Translation[2] (Literal)
kitsui	きつい	kee-t'sue-ee	difficult
kizuki	気付き	kee-zuu-kee	awareness … notice … realization
matomeru kanri	目で見る管理	mah-toe-may-ruu kahn-ree	visual management … management you can see
mieruka	見える化	mee-rue-kah	transformation into something visual … visualization
minomi	みのみ	me-no-me	main part
mirai	未来	me-rah-ee	the future
mizushimashi	水澄まし	me-zuu-she-mah-she	water beetle
mondai	問題	moan-dah-ee	problem … question
mondai kaiketsu	問題解決	moan-day-ee kye-kay-t'sue	problem-solving
monozukuri	物作り	moe-no-zuu-kuu-ree	hand made
muchi	無知	muu-chee	ignorance
muda	無駄	muu-dah	waste … futile, useless, idle, superfluous
mura	斑	muu-rah	uneven, irregular
muri	無理	muu-ree	overburdening … unreasonable … excessive
mushi	無視	muu-she	ignoring
muzukashii	難しい	muu-zuu-kah-shee	hard … difficult … troublesome
nagara	生産平滑化	nah-gah-rah	smoothing the flow of production
nemawashi	根回し	nay-mah-wah-she	lay a foundation … build consensus … behind-the-scenes talk … research
nichijou kanri	日常管理	nee-chee-joe-uu kahn-ree	management of the ordinary/regular
ningenseisoncho	人間性尊重	neen-gane-say-soan-choe	respect for humanity
obeya	大部屋	oh-bay-yah	great room … open office area
Ohno, Taiichi	大野耐一	tie-ee-chee oh-no	father of the Toyota Production System
oshiego	教え子	oh-shee-goe	student … disciple
poka yoke	ポカ避け	poe-kah yoe-kay	avoid inadvertent error … mistake/error proof

Term	Japanese (*Kanji*)	Pronunciation (Common)	English Translation[2] (Literal)
riso (or risou)	理想	ree-so	ideal … ideally
saitekika	最適化	sigh-tay-kee-kah	optimization
seiban	製番	say-bahn	product number
seicho	成長	say-choe	growth
seiketsu	清潔	say-kay-t'sue	standardize
seiri	整理	say-ree	sort … separate … declutter
seisan hoshiki	生産方式	say-sahn hoe-she-kee	manufacturing system
seiso	清掃	say-so	shine … cleanliness
seiton	整頓	say-tone	store … set in order … orderliness
sensei	先生	sin-say	teacher … professor … mentor
shain	社員	shah-een	employee
shido sha	指導者	she-doe shah	leader
shido (ryoku)	指導	she-doe (rio-kuu)	leadership
Shingo, Shigeo	新郷 重夫	she-ghee-oh sheen-go	Japanese consultant and developer of *Poka Yoke* and SMED
shitsuke	躾	she-t'sue-kay	sustain … discipline
shojinka	少人化	show-jeen-kah	flexible workforce/workplace
shuukai	集会	shuu-kye	meeting … assembly
shushinkoyo	終身雇用	shuu-sheen-koe-yoe	lifetime employment
sonkei suri	尊敬する	soan-kay sue-ree	respect
souzouryoku	創造力	so-uu-zoe-uu-rio-kuu	creative power … creativity
takumi	匠	tah-kuu-me	master craftsperson

THE ABIDIAN IIOT/CI DICTIONARY

Term	Japanese (*Kanji*)	Pronunciation (Common)	English Translation[2] (Literal)
tatakidai	叩き台	tah-tah-kee-dye	open discussion
tebanare	手離れ	teh-bah-nah-ray	hands-off
teiinsei	定足数	tay-ee-een-say	quorum system
tenchou	店長	tay-n-choe-uu	shop manager
toukan	統監	toe-uu-kahn	supervision … supervisor
Toyoda	豊田	toe-yoe-dah	surname of the founder(s) of Toyota Motor Corporation (note it is not "Toyota")
Toyoda, Kiichiro	豊田 喜一郎	kee-ee-chee-roe toe-yoe-dah	founder of what is today Toyota Motor Corporation
Toyoda, Sakichi	豊田 佐吉	sah-kee-chee toe-yoe-dah	loom inventor and founder of Toyota Industries … developed concept of *Jidoka*
tsukurikomi	作り込み	t'sue-kuu-ree-koe-me	manufacturing
yamazume (*yamazumi*)	山積み	yah-mah-zuu-may (yah-mah-zuu-me)	smoothing, leveling (a mound)
yokoten	横展開	yoe-koe-ten	spread (horizontal spread of an idea across an organization)
yuuryou	優良	yuu-rio-uu	superior, excellent, fine

[1] Editor's Note: I am very much a student of Japanese. My intent for learning the Japanese that I have is to gain a deeper understanding of the Toyota Production System. I believe an unbiased personal understanding can only be gained by discussing with and reading from the original masters … again, *genchi genbutsu*, if you will. ☺

I apologize in advance for those terms, *kanji*, pronunciations, and translations I have shown incorrectly. I humbly welcome corrections at info@abidian.com from those more experienced in the Japanese language.

[2] Refer to entries in the text for complete definitions as they apply to Continuous Improvement

About Abidian

Abidian, a global consulting firm, has been in business since 2005. We are known for our practical insight in delivering and implementing continuous improvement toolsets such as Lean, Six Sigma (6σ), Maintenance Excellence, Theory of Constraints (TOC), Statistical Process/Quality Control, and Total Productive Maintenance and Manufacturing (TPM). Over the years, Abidian's consultants have worked with such household names as ExxonMobil, Hewlett-Packard, Gillette, Honda, Milliken, Proctor & Gamble, Toyota, and the United States Air Force.

> Abidian is a professional services company dedicated to making organizations around the globe more productive. We direct our resources into coaching, providing services, and applying technologies that:
>
> - Stress the ***Practical*** application of the various continuous improvement tools,
> - Promote ***Acceptance*** and use at all levels of an organization,
> - Encourage a ***Culture*** of respect and *continuous* improvement, and
> - Exceed the greatest ***Expectations*** of our clients.
>
> At Abidian, we refer to this as helping our clients set a new ***P.A.C.E.***

Our consultants hold professional certifications and decades of practical, hands-on experience in the industries they serve. They have been recognized individually and collectively by entities such as the American Society for Quality, United States Congress, United States Commerce Association, the late Fred Thompson's *Inside Business Review* seen on CNN, and the late General Alexander Haig's *21st Century Business Review* seen on CNBC. In recognition of our system-related knowledge and expertise, Abidian has been recognized as one of the Top 10 Lean Manufacturing Consulting Companies.

From a software perspective, Abidian is proud to be a part of the Microsoft Partner

Network and a long-time member of Microsoft's Advisory Research Panel. Abidian's Microsoft representative, Mike Chambers, has over two decades experience working with Microsoft. Abidian is also a charter member of the Internet of Things Institute and a member of the Technical Advisory Group for the industry trade group MTConnect.

The bottom line is Abidian has the in-house expertise to create practical monitoring and analysis applications using cost-effective off-the-shelf products, like Microsoft Office and the hands-on experience to help clients get the most from their implementation. Mike Chambers' publication of *Real-Time Production Efficiency Monitoring with Microsoft Office* is indicative.

About the Editor

Mike Chambers is an experienced hands-on continuous improvement practitioner and a long-time member of the Microsoft Advisory Board. He is a Senior (Managing) Principal with Abidian, a global manufacturing consulting firm. With over 35 years of practical experience, Mike specializes in helping organizations achieve their highest potential by improving their culture while eliminating inefficiencies and ineffective work practices. He has held management and continuous improvement leadership responsibilities at both the facility and corporate levels with three Fortune 500 companies.

An electrical engineer by education, Mike is an accomplished Certified Lean Six Sigma Master Black Belt (CMBB). He is well-known for his down-to-earth, easily-understood explanations of complex concepts. Mike is also a hands-on Certified Maintenance and Reliability Professional (CMRP) and licensed master electrician. He has been independently recognized as a *sensei* for his working knowledge of Lean and the Toyota Production System.

Mike is the author of over two dozen books. When not writing, or helping others, Mike and his beautiful wife Kay enjoy hanging out on the Gulf Shores (Gulf Coast) beaches with their two neurotic cats.

THE ABIDIAN IIOT/CI DICTIONARY

Other Books by Mike Chambers

Abidian Continuous Improvement Lexicon: A Working Index of Keywords Associated with Lean, Six Sigma, Maintenance Excellence, Outside-In Thinking, Reliability, Statistics, Theory of Constraints, and Total Productive Maintenance and Manufacturing. Turn the Page Books, 2010.

Advanced Problem Solving: A Hands-On Approach to Three-Step Problem Solving. Abidian, 2010-2017.

Continuous Improvement in Schools with Lean Education and Smart Transition. Co-Author with Mary K. Chambers, Turn the Page Books, 2010.

Continuous Improvement in Schools with Lean Education and Strategic Planning. Co-Author with Mary K. Chambers, Turn the Page Books, 2010.

Failure Mode and Effects Analysis (FMEA). Abidian, 2005-2017.

First Light of the Truth. Self-Published, 2003.

Focusing Continuous Improvement and Lean Six Sigma Using Value Stream Mapping. Abidian, 2007.

Introduction to Lean and the Philosophy of Continuous Improvement (For Leaders). Abidian, 2010.

ISO 9001: 2000 Internal Auditor. Abidian, 2007.

Leadership 101: Changing a Culture. Abidian, 2011-2017.

Lean Education. Abidian Institute, 2005-2017.

Lean Maintenance/Maintenance Excellence. Abidian, 2005-2017.

Lean Manufacturing Execution: A Practical Approach to Long-Term Success with Lean. Abidian, 2005-2017.

Lean Manufacturing/Operating Excellence. Abidian, 2008.

Lean Office. Abidian Institute, 2005-2017.

Moorer-Murer Kinfolks. Self-Published, 1989.

Real-Time Production Efficiency Monitoring with Microsoft® Office: A Practical Approach to Applying Overall Equipment Effectiveness and the Industrial Internet of Things in a Lean Manufacturing Workplace. Austin Cove Books, 2017.

Six Sigma Green Belt: A Hands-On Emphasis of the Practical Aspects and Implementation of Six Sigma for Continuous Improvement Project Leaders. Abidian, 2005-2017.

Six Sigma Yellow Belt: Introducing Six Sigma. Abidian, 2016.

Statistical Process Control (SPC). Abidian, 2007.

Strategic Lean Leadership: Introducing a New Way to Lead. Abidian, 2014-2017.

Total Productive Manufacturing: A Practical Approach to Achieving Manufacturing Excellence. Abidian, 2005-2017.

Using SmartThings and Amazon Echo to Simplify, Enhance, and Secure: A Practical, Cost-Effective Approach to bringing the IoT (Internet of Things) to Home and Business Automation. Expected 2017 publication.

Index

0

0-9 · 3

1

1.5 Sigma Shift · 3
14 Management Principles (of Lean/The Toyota Way) · 3
14 Points · *See* See Deming's 14 Points

2

2.4 GHz · 3, 5

3

3 "G" Principles · 3
3 Elements of Demand · 3, 95, 107
3 Ms · 77, 78, 122
3 Principles of Lean · 4, 84, 92, 94, 111
3 Step Change Process · 4
3"G" Principles · 48
3D Printing · *See* Additive Manufacturing, *See* Additive Manufacturing
3Ds · 4
3Ks · *See* 3Ds
3Ms · 4

3P · 4
3Rs · 4

4

4 Ms · *See* Four Ms
4G · 4, 5, 70
4Ms · 5, 46
4P Model · 4
4th Industrial Revolution · 4, 56

5

5 GHz · 5
5 Whys · 5, 100
5G · 4, 5
5Ms · *See* 5Ms of Production
5Ms of Production · 4, 5
5S · 5, 6, 44, 98
5S Audit · 5
5S Cart · 5
5S Scan · 5
5S+1 · 5, 6
5W's and 1H · 5

6

6 Sigma · *See* Six Sigma (6σ)
6S · *See* 5S
6σ · *See* Six Sigma (6σ)

THE ABIDIAN IIOT/CI DICTIONARY

7

7 Flows · 6
7 Tools of SPC/SQC · 6
7 Wastes · 6, 84, 86

8

8 Wastes · 6, 9, 20, 28, 34, 40, 59, 63, 68, 76, 77, 86, 113, 115, 125
80/20 Rule · 6, 88, 124
802.11 · *See* IEEE 802.11
8D Corrective Action Process · 6

A

A · 7
A3 · 7
A3 Problem Solving · 7, 93
ABC · 7
Abidian · 141
Abnormality · 7, 11
Abnormality Detection · 11
Abnormality Management · 7, 56, 123, 137
AC · *See* Alternating Current (AC)
Accelerometer · 7
Acceptable Quality Level (AQL) · 8, 12
Access Control · 8
Access Control as a Service · 8
Access Point (AP) · 8
Accuracy · 8
Action Plan · 8
Activity Based Costing (ABC) · 7, 8
Activity Board · 8
Activity Network Diagram · 8, 94
Activity Sheet · 8
Activity Tracker · 8
Actuator · 8, 115
ADAS · *See* Advanced Driver Assistance System (ADAS)
ADC · *See* Analog-to-Digital Converter (ADC)
Additive Manufacturing · 4, 9
Addressability · 9
Administrative Processes · 123
Administrative/Service Wastes · 9
Advanced Driver Assistance System (ADAS) · 9
Advanced Equipment Standard (AES) · 9
Advanced Message Queuing Protocol (AMQP) · 9, 10

Advanced Product Quality Planning (APQP) · 9, 12
AES · *See* Advanced Equipment Standard (AES)
Aesthetics · 9
Affinity Diagram · 9
Agile Management · 9
AI · *See* Artificial Intelligence (AI)
Alexa · 10
Alpha (α) · 10
AltBeacon · 10
Alternating Current (AC) · 7, 10
Amazon Echo · 10, 146
Amazon Web Services (AWS) · 10, 14
Ambient Intelligence (AmI) · 10
AmI · *See* Ambient Intelligence (AmI)
Amps/Amperes · 10
AMQP · *See* Advanced Message Queuing Protocol (AMQP)
Analog · 10, 36
Analog-to-Digital Converter (ADC) · 9, 10
Analysis of Means (ANOM) · 10
Analysis of Variance (ANOVA) · 11, 43, 58, 74, 84, 102
Analyze · 11, 37
Anderson-Darling Normality Test · 11
Andon · 11, 90, 123, 136
Andon Cord · 11
Android · 11
Android Device · 19
Android Wear · 11
ANN · *See* Artificial Neural Network (ANN)
Annual Inventory (Stock) Turns · 59
Anode · 11
Anomaly Detection · 7, 11
ANOVA · 11, 43
Anthropometry · 12, 41
AP · *See* Access Point (AP)
API · *See* Application Program Interface (API)
Application Program Interface (API) · 12
Appraisal Costs · 12
APQP · *See* Advanced Product Quality Planning (APQP)
AQL · *See* Acceptable Quality Level (AQL)
Arduino · 12
Arithmetic Mean · 12
Arrow Diagram · 12
Artificial Intelligence (AI) · 10, 12, 24
Artificial Neural Network (ANN) · 11, 12, 80
AS9100 · 12
Ashton, Kevin (1968-) · 13

Assembly Line · 13, 73
Asset Management · 13
Asset Tracking · 13
Assignable Causes (of Variation) · 13
Attribute · 13
Attribute Chart · 13
Attribute Data · 13, 36
Attribute Nominal Data · 13
Attribute Ordinal Data · 13
Auto Time · 14
Automatic Time · 14, *See* Auto Time
Automation · 13
Autonomation · 14, 61, 114, 137
Autonomous Automation · *See* Autonomation
Autonomous Maintenance · 14, 84, *See* Online Checks
Availability · 14, 86, 118, 119
Average · 12, 14

B

B · 15
Backflow · 15
Backlog · 15
Baka Yoke · 136
Balanced · 15
Balanced Scorecard · 15
BAN · *See* Body Area Network (BAN)
Band · 15
Bandwidth · 15
Bar Chart · 15, 52, 88
Bar Graph · 15, 129
Barrier Analysis · 100
Base 16 · *See* Hexidecimal
Baseline · 16
Batch and Queue · *See* Batch Operation and Mass Production
Batch Operation · 16, 73, 94, 114
Batch Size · 16
Bathtub Curve · 16, 126, *See* Weibull Distribution
Baud · 16
Bayes Theorem · 16
Bayesian Inference · 100
Beacon · 10, 16, 39
Bell Curve · 16, 24, 41
Benchmarking · 16
Benefit-to-Cost Ratio · 16
Best Practice · 17, 108
Beta (β) · 17
BI · *See* Business Intelligence (BI)

Bias · 101
Bias (in Measurement) · 17
Big Data · 17, 51
Binary · 17, 18, 21
Binomial Data · 36
Binomial Distribution · 17
Bisertial Correlation Coefficient · 17
Bit · 17, 21
Black Belt (BB) · 17, 49, 73, 104, 126, 131
Blast · 17
Blitz · *See kaizen*
Block Diagram · 18
Bluetooth · 3, 18
Bluetooth Low Energy · 18, 19
BMS · *See* Building Management System (BMS)
Body Area Network (BAN) · 15, 18
Boolean · 18
Bootloader · 18
Botnet · 18, 76
Bottleneck · 18, 27, 76, 112
Bounce · 18
Bowling Chart · 18
Box Plot · 18, 126
BPE · 18
BPM · *See* Business Process Management (BPM)
BPMN · *See* Business Process Model and Notation (BPMN)
Brainstorming · 19
Breakdown · 19
Breakthrough · 19
Brick · 19
Brillo · 19
Bring Your Own Device (BYOD) · 19, 29
Broker Server · 19
Brownfield · 19
Browser · 19
Bubble Diagram · 19
Bucket Brigade · 20
Budgeted Production Efficiency (BPE) · 20, 86
Buffer Stock · 20
Building Management System (BMS) · 18, 20
Build-to-Order · 20
Built-in-Quality · 20
Bull-Whip Effect · 20, 34
Burabura · 136
Business Intelligence (BI) · 17, 20
Business Process Management (BPM) · 19, 20
Business Process Model and Notation (BPMN) · 19, 20
Business Value Added (BVA) · 21, 80, 121
BusyBox · 21

THE ABIDIAN IIOT/CI DICTIONARY

BYOD · *See* Bring Your Own Device (BYOD)
Byte · 21

C

C · 23
C of V · 23
C&E Diagram · *See* Cause and Effect (Fishbone) Diagram
C/O · *See* Changeover (C/O)
CA · *See* Certificate Authority (CA)
CAP · *See* Constrained Application Protocol (CAP)
Capability · 23, 29
Cargo Cult Science · 23, 44
Catchball · 23
Cathode · 23
Causal Factor Tree Analysis · 100
Causal Flow Diagram (CFD) · 23
Cause and Effect (Fishbone) Diagram · 6, 23, 24, 44, 60, 100
Cause and Effect Matrix · 24
CBM · 24
CC · *See* Creative Commons (CC)
c-Chart · 24
CCPM · *See* Critical Chain Project Management (CCPM)
Cell · 24, 25, 127
Cellular Manufacturing · 24, 25, 117, 127
Centaur Model · 24
Central Limit Theorem · 16, 24
Central Processing Unit (CPU) · 24, 30
Certificate Authority (CA) · 23, 25
CFD · *See* Causal Flow Diagram
Chaku Chaku · 25, 51, 136
Chalk Circle Exercise · 25
Chance Variation · 25
Change Agent · 25
Change Analysis · 100
Change for the Worse · 137
Changeover (C/O) · 23, 25
Charter · 8, 25, 65
Chebyshev's Theorem · 25
Check Sheet · 6, 25
Chi-Square Distribution · 25, 49
Choko choko tomari · 136
CI · *See* Continuous Improvement (CI)
Circuit · 26
Client · 26, 103
Cloud Computing · 26
CMMS · *See* Computerized Maintenance Management System (CMMS)
Coaching *Kata* · *See Kata*
Coefficient of Determination · 26
Coefficient of Variation · 23, 26
Cognitive IoT · 26
Cognitive Radio · 26
COGS · *See* Cost of Goods Sold
Cold Chain · 26
Comfort Design · 41
Common Cause Variation · 26, 107, 122
Companion Device · 27
Complement of an Event · 27
Compound Event · 27
Computerized Maintenance Management System (CMMS) · 26, 27
Condition Based Maintenance (CBM) · 27
Condition Based Monitoring (CBM) · 24, 27, 84
Conditional Probability · 27
Conductor · 27
Confidence Level · 27
Constrained Application Protocol (CAP) · 23, 27
Constraint · 18, 27, 112
Consumer · *See* Customer
Continuous Data · 28, 101
Continuous Flow · 28
Continuous Improvement · 145
Continuous Improvement (CI) · 4, 8, 25, 26, 27, 65, 68, 85, 135, 136
Continuous Initiatives · 28, 44
Continuous Variable · *See* Continuous Data
Control · 28, 37
Control Charts · 6, 24, 28, 100, 107, 117
Controller · 28
Controls · 28
Conveyance · 28
CONWIP · 28
COPE · *See* Corporate-Owned, Personally-Enabled (COPE)
COPQ · *See* Cost of Poor Quality (COPQ)
Corporate-Owned, Personally-Enabled (COPE) · 19, 28, 29
Corrective Action · 6, 29
Correlation · 29, 58, 102
Correlation and Regression Analysis · 102
Correlation Coefficient · 29
Cost of Goods Sold (COGS) · 26, 29, 59
Cost of Poor Quality (COPQ) · 28, 29
Count of Items · 29
Counter-Clockwise Flow · 29
C_p · 23, 29

C_{pk} · 23, 29
CPPS · *See* Cyber-Physical Production System (CPPS)
CPS · *See* Cyber-Physical System (CPS)
CPU · *See* Central Processing Unit (CPU)
Creative Commons (CC) · 24, 30
Creative Commons Attribution – ShareAlike · 30
Critical Chain Project Management (CCPM) · 24
Critical Failure · 30
Critical Failure Mode · 30
Critical to Quality/Customer (CTQ or CTC) · 30
Critical Value · 30
CRM · *See* Customer Relationship Manager (CRM)
Crosby, Philip B. (1926-2001) · 30, 133
Cross-Platform · 30
Crowdfunding · 30
CryptoChip · 30
C_T · *See* Cycle Time (C_T)
CTQ or CTC · *See* Critical to Quality/Customer (CTQ or CTC)
Culture · *See* Lean Culture
Cumulative Distribution Function · 31
Current · 31
Current State Map · 31, 46, 122
Customer · 31
Customer Demand · 31, 34
Customer Relationship Manager (CRM) · 30, 31
Customer Segmentation · 31
Customer-Centric · 85, 123
Cyber-Physical Production System (CPPS) · 29, 31
Cyber-Physical System (CPS) · 29, 31
Cycle Time (C_T) · 30, 31, 69, 111

D

D · 33
D of F · *See* Degrees of Freedom (D of F)
DAC · *See* Digital-to-Analog Converter (DAC)
Dangerous · 4, 137
Dashboard · 33
Data Center · 33
Data Logger · 33
Data Mining · 33
Data Scientist · 33
Data Set · 33
Data Structure · 60
Data-Driven Decision Management (DDDM) · 33
Datakinesis · 33
DC · *See* Direct Current (DC)
DDDM · *See* Data-Driven Decision Management (DDDM)
Debug · 34
Decision/Event Tree · 34
Defective · *See* Non-Conformance
Defects · 34, 80, 99, 113
Defects Per Million Opportunities (DPMO) · 34, 37, 92, 104
Defects Per Unit (DPU) · 34, 37
Define · 34, 37
Degrees of Freedom (D of F) · 33, 34
De-Identification · 34
Delivery Performance · 34
Demand · *See* Customer Demand
Demand Amplification · 20, 34
Demand Response · 35, 69
Deming Cycle · 88, *See* PDCA
Deming, W. Edwards (1900-1993) · 35, 109, 114, 118
Deming's 14 Points · 3, 35
Dependent Samples · 35
Dependent Variable · 26
Descriptive Statistics · 35
Design · 37
Design for Maintainability · 35
Design for Manufacturing (DFM) · 35, 36
Design for Six Sigma (DFSS) · 35, 36
Design of Experiment (DOE) · 35, 37
Device · 35
    Intelligent · 58
Device Attack · 36
Devil's Advocacy · 36
DFM · *See* Design for Manufacturing (DFM)
DFSS · *See* Design for Six Sigma (DFSS)
Difficult · 4, 138
Digital · 10, 36
Digital-to-Analog Converter (DAC) · 33, 36
Diode · 11, 23, 36
Direct Current (DC) · 34, 36
Direct Memory Access (DMA) · 36, 37
Dirty · 4, 137
Discrepancy · *See* Non-Conformance
Discrete Data · 13, 36, 101
Discrete Variable · 36
Distribution · 36
Divergent Thinking · 36
DIY · *See* Do It Yourself (DIY)

DL or D/L · *See* Download (DL or D/L), *See* Downlink (DL or D/L)
DMA · *See* Direct Memory Access (DMA)
DMADV · 37
DMAIC · 11, 17, 28, 34, 37, 56, 74
DNP3 Protocol · 37
Do It Yourself (DIY) · 37
DOE · *See* Design of Experiment
Downlink (DL or D/L) · 37
Download (DL or D/L) · 37
Downtime · 37
DPMO · *See* Defects Per Million Opportunities (DPMO)
DPU · *See* Defects Per Unit
Drum-Buffer-Rope · 37

Error Proofing · 45, *See Poka Yoke*
ESD · *See* Electrostatic Discharge (ESD)
Ethernet · 41
European Association of Aerospace Industries · 12
Event · 41
Excess Processing · *See* Over Processing
Excess WIP · 42
Executive Standard Work · 41, 67
Expert System · 42
Express Train · 42
Extended Team Members · 42
Extensible Markup Language (XML) · 42, 129
Extensible Messaging and Presence Protocol (XMPP) · 42, 129, 130
External Setup · 42

E

E · 39
Economic Order Quantity (EOQ) · 39, 41
Economy of Scale · 39
ECRS · *See* Eliminate, Combine, Rearrange, and Simplify (ECRS)
Eddystone · 39
Edge Computing · 39
Edge Gateway · 39
Edison · 39
EDR · *See* Electrodermal Response (EDR)
EEPROM · 39
Effectiveness · 40
Efficiency · 40
Eight Wastes · *See* 8 Wastes
Electrodermal Response (EDR) · 39, 40, 47
Electrostatic Discharge (ESD) · 40, 41
Elements of Work · 40
Eliminate, Combine, Rearrange, and Simplify (ECRS) · 39, 40
Embedded System · 40
Embrace, Extend, and Extinguish · 40
Emergency Stock · *See* Safety Stock
Empirical Data · 40
Empirical Probability · 40
Empirical Rule · 41
Employee · 139
Enernet · 41
Enterprise Resource Planning (ERP) · 41, 73
Environmental Management Systems · 60
EOQ · *See* Economic Order Quantity (EOQ)
Equipment Location Board · 41
Ergonomics · 12, 41
ERP · *See* Enterprise Resource Planning (ERP)

F

F · 43
f Distribution · 43
Factor · 43
FACTUAL · 43
Fads · *See* Continuous Initiatives
Failure · 43
Failure Effect · 43
Failure Mode · 43
Failure Mode Analysis (FMA) · 43
Failure Mode and Effects Analysis (FMEA) · 44, 45, 100, 145
Failure Rate (λ) · 44
Fake Lean · 23, 44
FIFO · *See* First-In First-Out
Finished Goods · 44
Firefighting · 100
Firmware · 44
Firmware Over-the-Air (FOTA) · 44, 46
First Pass Yield · *See* First Time Yield (FTY)
First Time Yield (FTY) · 20, 44
First-In First-Out (FIFO) · 44, 67
Fishbone Diagram · *See* Cause and Effect (Fishbone) Diagram
Five S · *See* 5S
Flavor-of-the-Month · *See* Continuous Initiatives
Flexible Workforce · 139
Flow · 45
Flow Chart · 6, 45
Flow Production · 45
Flow Shop · 45
Flowchart · 109

FMA · *See* Failure Mode Analysis (FMA)
FMEA · *See* Failure Mode and Effects Analysis (FMEA)
Focus Group · 45
Focused Improvement · 45
Fog Computing · 45
Fog Network · *See* Fog Computing
Fogging · *See* Fog Computing
Fool Proofing · 45
Ford Motor Company · 45
Ford, Henry (1863-1947) · 45
Form · 137
FOTA · *See* Firmware Over-the-Air (FOTA)
Four Ms · *See* 4Ms
Franklin, Benjamin (1706-1790) · 46
Frequency Diagram · *See* Histogram
Frequency Distribution · 46
Frequent Conveyance · 46
f-Test · 43
Fuchsia · 46
Functional Design · 41
Functional Failure · *See* Failure
Futekisetsu na · 136
Future State Map · 46, 55, 122

G

G · 47
Gage R&R · 47, 99
Galvanic Response · *See* Electrodermal Response (EDR)
Gantt Chart · 47, 94
Gap Analysis · 47
Gates, Bill (1955-   ) · 47
Gateway · 47
Gauss, Johann Carl Friedrich/Frederick (1777-1855) · 47
Gaussian Curve · 47
GB · *See* Green Belt (GB)
Gemba · 3, 48, 72, 136
Gemba Walk · 48
Gembutsu · 48
Genbutsu · 3, 48, 136
Genchi · 136
Genchi Genbutshu · 136
Genchi Genbutsu · 3, 48, 72, 127
General Motors · 80
Genjitsu · 3, 48, 136
Geofence · 48
Geographic Information System (GIS) · 48
Gilbreth, Frank (1868-1924) · 48, 112, *See*

Therblig Analysis
GIS · *See* Geographic Information System (GIS)
Global Positioning System (GPS) · 48, 49
Go and See Management · *See Genchi Genbutsu*
Goals · *See* Vision
Goldratt, Eliyahu (1947-2011) · 48, 62, 112, *See* Theory of Constraints (TOC)
Good Thinking, Good Products · 49
Goodness-of-Fit Test · 49
Gourika · 136
GPS · *See* Global Positioning System (GPS)
Graphical User Interface (GUI) · 49
Great Room · 138
Green Belt (GB) · 47, 49, 126, 131, 146
Green Washing · 49
Greenfield · 49
GUI · *See* Graphical User Interface (GUI)

H

H · 51
Hadoop · 51
Hakarigoto · 136
HaLow · 51
Hanashiai Toron · 136
Hancho · 136
Hanedashi · 51, 136
Hansei · 51, 136
Haptic Technology · 51
Haptics · *See* Haptic Technology
Harada Method · 51
Hawthorne Effect · 52
HAZOP · 52
Heijunka · 52, 68, 94, 106, 136
Heijunka Box · 52
HEM · *See* Home Energy Management (HEM)
Hertzian Space · 52
Hex · *See* Hexidecimal
Hexadecimal · 52
Hi kōritsuteki · 136
Hidden Factory · 15
Hinshitsu · 136
HIPPO · 52
Histogram · 6, 52
Hitozukuri · 137
HMI · *See* Human Machine Interface (HMI)
Home Energy Management (HEM) · 52, 53
Hoshin · 53, 91
Hoshin Kanri · 91, 137, *See Hoshin*
Hoshin Planning · *See Hoshin Kanri*

Host · 53
House of Quality · 53
Hub · 53
Human Machine Interface (HMI) · 49, 52, 53, 71, 98
Human Optimization · 53
HVAC · 53
Hypergeometric Distribution · 53
Hypothesis Tests · 53

I

I · 55
I/O · *See* Input/Output (I/O or IO)
IaaS · *See* Infrastructure as a Service (IaaS)
iBeacon · 55
IC · *See* Integrated Circuit (IC)
Ichidan · 137
ICS · *See* Industrial Control System (ICS)
IDE · *See* Integrated Development Environment (IDE)
Ideal · 115
Ideal State Map · *See* Future State Map
Ideals · *See* Vision
IEC · 55, 60
IEEE · 55
IEEE 802.11 · 6, 51, 56
IFTTT · 56
IGES · *See* Initial Graphics Exchange Specification (IGES)
IGMP · *See* Internet Group Management Protocol (IGMP)
IIoT · *See* Industrial Internet of Things (IIoT)
IISF · *See* Industrial Internet Security Framework (IISF)
Iji · 137
Ijo Kanri · 7, 56, 137
Improve · 37, 56
Improvement *Kata* · *See Kata*
Independent Variable · 26
Indoor Positioning System (IPS) · 56, 59
Industrial Control System (ICS) · 55, 56
Industrial Internet of Things (IIoT) · 1, 2, 4, 39, 56, 58, 75, 106, 146
Industrial Internet Security Framework (IISF) · 56
Industry 4.0 · *See* 4th Industrial Revolution
Infant Mortality · 57
Inferential Statistics · 57
Information Flow · 57
Information Technology (IT) · 56, 57, 60

Infrastructure as a Service (IaaS) · 55, 57
Initial Graphics Exchange Specification (IGES) · 56, 57
Input · 57
Input X's · 129, *See* X Data/Values
Input/Output (I/O or IO) · 55, 57, 59
Inspection · 57
Installer · 57
Insulator · 57
Insurance Telematics · 57
Integrated Circuit (IC) · 55, 58
Integrated Development Environment (IDE) · 55, 58
Intelligent Device · 58, 105, 106
Intelligent Transport System (ITS) · 58, 60
Interaction · 58
Interaction Effect · 58
Interdependence Test · 58
Interdependent Variable · 58
Interim Containment Action (ICA) · 6
Internal Mobile Subscriber Identity (IMSI) · *See* Subscriber Identity Module (SIM)
Internal Setup · 42, 58
International Electrotechnical Commission (IEC) · 55
International Telecommunications Union (ITU) · 58, 60
Internet · 58, 59, 123
Internet Group Management Protocol (IGMP) · 56, 58
Internet of Things (IoT) · 19, 26, 27, 39, 58, 59, 60, 76, 89, 125, 129
Internet Protocol (IP) · 58, 59
Interrelationship Digraph · 59
Intranet · 59, 123
Introduction · 1
Inventory · 20, 59, 63, 113
Inventory Turns · 59, 115
IO · *See* Input/Output (I/O or IO)
IoT · *See* Internet of Things (IoT)
IP · *See* Internet Protocol (IP)
IP Address · 59
IPS · *See* Indoor Positioning System (IPS)
iPv4 · 59
iPv6 · 59
Irregular · 138
Ishikawa Diagram · *See* Cause and Effect (Fishbone) Diagram
Ishikawa, Kaoru (1915-1989) · 24, 60, 96, 137
ISO · 60
ISO 13053 · 60
ISO 14001 · 60

ISO 17258 · 60
ISO 18404 · 60
ISO 9001 · 60, 145
ISO Quality Management Systems · 1
ISO/IEC 29161 · 60
IT · *See* Information Technology (IT)
ITS · *See* Intelligent Transport System (ITS)
ITU · *See* International Telecommunications Union (ITU)

J

J · 61
Japanese Terms · 135
JavaScript Object Notation (JSON) · 61, 62
JCAHO · *See* Joint Committee for the Accreditation of Healthcare Organizations (JCAHO)
JerryScript · 61
JI · *See* Job Instruction (JI)
Jidoka · 61, 114, 137, 140
Jishuken · 61, 137
JIT · *See* Just in Time
JM · *See* Job Methods (JM)
Job Instruction (JI) · 61, 62
Job Methods (JM) · 62
Job Relations (JR) · 62
Job Shop · 62
Joint Committee for the Accreditation of Healthcare Organizations (JCAHO) · 61, 62
Jonah · 62
Jones, Daniel T. · 62, 127
JR · *See* Job Relations (JR)
JSON · *See* JavaScript Object Notation (JSON)
Juran Trilogy · 62
Juran, Joseph M. (1904-2008) · 62, 124
JUSE · *See* Union of Japanese Scientists and Engineers
Just in Sequence · 63
Just in Time (JIT) · 45, 63, 66, 109, 137
Just in Time Training · 63
Juugyouin · 137

K

K · 65
Kaiaku · 65, 137
Kaikaku · 65, 137
Kaizen · 3, 8, 41, 61, 65, 91, 135, 137
Kaizen Event · 17, 65, 91

Kakushin · 137
Kalman Filter · 65
Kamishibai · 137
Kamishibai Board · 65, 111
Kanban · 52, 66, 75, 109, 137
Kanban Hōshiki · 137
Kangae · 137
Kano Analysis · 66
Kano Model · 66
Karakuri · 137
Karoshi · 66, 137
Kata · 66, 137
Katashiki · 137
Katashiki Card · 66
Keiretsu · 137
Kepner-Tregoe · 100
Key Performance Indicator (KPI) · 18, 66, 75
Kikanshi · 137
Kiken · 4, 137
Kingman's Formula · 66
Kitanai · 4, 137
Kitsui · 4, 138
Kitting · 66
Kizuki · 138
KPI · *See* Key Performance Indicator (KPI)
Kruskal-Wallis Test · 66

L

L · 67
Labor Linearity · 67
LAMBA · 67
LAN · *See* Local Area Network (LAN)
Last-In First-Out (LIFO) · 44, 67, 69
LCC · *See* Life Cycle Cost
LCL · *See* Lower Control Limit
Lead Time · 67, 69, 73, 93, 96
Leader · 139
Leader Standard Work (LSW) · 41, 67, 70, 108
Leadership · 139, 146
Lean · 1, 4, 5, 23, 44, 45, 46, 48, 60, 62, 65, 68, 92, 93, 94, 103, 108, 111, 121, 141, 145, 146
Fake · 23
Lean Culture · 31, 68
Lean Education · 146
Lean Enterprise · 68
Lean Enterprise Institute · 127
Lean Maintenance · 146
Lean Manufacturing · 146
Lean Office · 146
Lean Principles · *See* Principles of Lean

Lean Six Sigma · 49, 73, 131
LED · *See* Light Emitting Diode (LED)
Left Hand/Right Hand Analysis · 68
Level of Significance · 68
Level Production · *See Heijunka*
Level Scheduling · 68, 106
Level Selling · 68
Leveling · 136
Library · 69
Life Cycle · 113
Life Cycle Cost (LCC) · 67, 69
Lifetime Employment · 139
Li-Fi · *See* Light Fidelity (Li-Fi)
LIFO · *See* Last-In First-Out (LIFO)
Light Emitting Diode (LED) · 68, 69
Light Fidelity (Li-Fi) · 69, 123
Line Balancing · 69
Linux · 69, 131
LiPo Battery · 69
Little's Law · 69
Load · 69
Load Shedding · 35, 69
Local Area Network (LAN) · 41, 67, 70, 123, 126
Long Range (LoRa) · 70
Long Term Evolution (LTE) · 70
LoRa · *See* Long Range (LoRa)
Lot Size · 70
Lower Control Limit (LCL) · 67, 70
LSW · *See* Leader Standard Work (LSW)
LTE · *See* Long Term Evolution (LTE)

M

M · 71
M2M · *See* Machine-to-Machine (M2M)
M2P · *See* Machine-to-Person (M2P)
MAC Address · 71
MAC/Macintosh · 71
Machine Capacity · 71
Machine Data · 71
Machine Interface · 71, 98
Machine That Changed the World, The · 62, 68, 127
Machine Time · 71, 72
Machine Work · 71, 72, 73
Machine-to-Machine (M2M) · 71, 72, 75
Machine-to-Person (M2P) · 71, 72
Main Effect · 72
Maintainability · 72
Maintenance · 72
Maintenance Excellence · 1, 72, 141, 146

Make-to-Order (MTO) · 72, 77, *See* Build-to-Order (BTO)
Make-to-Stock (MTS) · 72, 77
Malware · 72, 76
Manage by Objectives (MBO) · 72, 74
Management by Wandering Around (MBWA) · 48, 72, 74
Manual Time · 71, 72, 73
Manual Work · 72, 73
Manufacturing Cycle Efficiency (MCE) · 73, 74
Manufacturing Standards · 77
Manufacturing System · 139
Manufacturing Technology · 73
Mass Production · 16, 45, 73, 94, 114
Master Black Belt (MBB) · 73, 104, 126
Material and Information Flow Diagram (MIFO) · 73, 75
Material Flow · 57, 73
Material Requirements Planning (MRP) · 41, 73, 77
Matrix Diagram · 73
MBB · 73
MBO · *See* Manage by Objectives (MBO)
MBWA · *See* Management by Wandering Around (MBWA)
MCE · *See* Manufacturing Cycle Efficiency (MCE)
MCU · *See* Microcontroller Unit (MCU)
Mean · 12, 14, 26, 35, 70, 74, 87, 91, 104
Mean Square · 74
Mean Time Between Failures (MTBF) · 74
Mean Time To Repair (MTTR) · 74
Measure · 37, 74
Measurement Scale · 74
Measurement System · 47
Measurement System Analysis (MSA) · 74
Mechatronics · 74
Medemiru Kanri · 138
Media Access Control (MAC) Address · *See* MAC Address
Median · 75, 126
Mesh · 75
Message Oriented Middleware (MOM) · 75, 76
Message Queuing Telemetry Transport (MQTT) · 75, 77
Metric · 75
Microcontroller Unit (MCU) · 74, 75
Midrange · 75
Mieruka · 138
MIFO · *See* Material and Information Flow Diagram (MIFO)
Milk Run · 75

MIMO · *See* Multiple-Input and Multiple-Output (MIMO)
Min/Max Inventory · 75
Mind Map · 75
Minomi · 76, 138
Minor Stop · 76
Mirai · 76, 138
Mistake Proofing · 45, 76, 95, 103
Mizushimashi · 138
Mobile Personal Emergency Response System (mPERS) · 76
Modal Class · 76
Modbus · 76
Mode · 76
MOM · *See* Message Oriented Middleware (MOM)
Mondai · 138
Mondai Kaiketsu · 138
Monozukuri · 138
Monte Carlo Simulation · 76
Monument · 18, 76, 87
Motion · 68, 76, 106, 112, 113
Motor · 77
mPERS · *See* Mobile Personal Emergency Response System (mPERS)
MQTT · *See* Message Queuing Telemetry Transport (MQTT)
MRP · *See* Material Requirements Planning (MRP)
MTConnect · 77
MTO · *See* Make-to-Order (MTO)
MTS · *See* Make-to-Stock (MTS)
Muchi · 138
Muda · 4, 77, 138
Multi-Machine Handling · 77
Multinomial Distribution · 77
Multiple-Input and Multiple-Output (MIMO) · 75, 77
Multi-Skilled · 77
Multi-Voting · 77
Mura · 4, 78, 138
Muri · 4, 66, 78, 138
Mushi · 138
Mutually Exclusive Events · 78
Muzukashii · 138

N

N · 79
Nagara · 138
Nagara Switch · 79

Nagara System · 79
Natural Team · 79, 127
Near Field Communications (NFC) · 79, 80
Nearable · 79
Negative Relationship · 79
Negawatt · 79
Nemawashi · 80, 138
Net Present Value (NPV) · 80
Neural Network (NN) · 12, 80
New United Motor Manufacturing, Inc. (NUMMI) · 80, 81
NFC · *See* Near Field Communications (NFC)
Nichijou Kanri · 138
Ningenseisoncho · 138
NN · *See* Neural Network (NN)
Node · 8, 80
Noise · 80
Nominal Data · 36
Non-Conformance · 34, 80
Non-Destructive Testing (NDT) · 80
Nonparametric Statistics · 80
Non-Value Added Activities · 21, 40, 80, 121
Non-Value Added Time · 113
Normal Curve · 80, 81
Normal Distribution · 81
Null Hypothesis · 81, 126
NUMMI · *See* New United Motor Manufacturing, Inc. (NUMMI)
NVA · 81

O

O · 83
OAE · *See* Overall Asset Efficiency (OAE)
Obeya · 83, 138
OCE · *See* Overall Classroom Effectiveness (OCE)
OEE · *See* Overall Equipment Effectiveness
Ogive · 83
Ohm · 83
Ohm's Law · 83, 99, 124
Ohno, *Taiichi* (1912-1990) · 6, 25, 84, 138
OLAP · *See* Online Analytical Processing (OLAP)
OMCD · *See* Operations Management Consulting Division (OMCD)
One Piece Flow · 4, 16, 45, 84
One Point Lesson (OPL) · 84, 104
One Touch Exchange of Die (OTED) · 84, 85
One-Tailed Test · 84
One-Way ANOVA · 84

Online Analytical Processing (OLAP) · 84
Online Checks · 14, 27, 84
OODA · 84
Open Room Effect · *See* Obeya
Open Source · 84
Open-Ended Distribution · 85
Open-Source · 11
Operating Excellence · 85, 146
Operating System (OS) · 11, 19, 30, 85, 90, 118, 127
Operational Definitions · 85
Operations Management Consulting Division (OMCD) · 83, 85
OPL · *See* One Point Lesson
Ordinal Data · 85
OS · *See* Operating System (OS)
OSE · *See* Overall Service Efficiency
Oshiego · 138
OTED · *See* One Touch Exchange of Die (OTED)
Outlier · 85
Output · 85
Output Y's · 129, 131, *See* Y Data/Values
Outside-In Thinking · 45, 85, 123
Over Processing · 86, 113
Over Production · 86, 113
Overall Asset Efficiency (OAE) · 83, 86
Overall Classroom Effectiveness (OCE) · 83, 86
Overall Equipment Effectiveness (OEE) · 14, 20, 83, 86, 146
Overall Labor Efficiency (OLE) · *See* Overall Service Efficiency (OSE)
Overall People Efficiency (OPE) · *See* Overall Service Efficiency (OSE)
Overall Service Effectiveness (OSE) · 86
Overburdening · 138

P

P · 87
Pacemaker · 76, 87
Pack Out Rate · 90, *See* Pitch
Packet Switching · 59, 87
PAN · *See* Personal Area Network (PAN)
Parallel · 87
Parametric Tests · 87
Pareto Analysis · 52, 88, 100
Pareto Chart · 88
Pareto Diagram · 6
Pareto Principle · 6, 88, 124
Pareto, Vilfredo (1848-1923) · 88
Parkinson's Law · 88
PCE · *See* Process Cycle Efficiency (PCE)
p-Chart · 24, 87
PD · *See* Program Development (PD)
PDCA · 8, 35, 67, 88, 103
PDCA Problem Solving · 88
PDF · *See* Probability Density Function (PDF)
PdM · 88
PDPC · *See* Process Decision Program Chart
PDSA · 88
Pearson Correlation Coefficient · 88
PEC · *See* Process Evaluation Checklist (PEC)
Performance · 86, 89
Peripheral · 89
Permanent Corrective Action (PCA) · 6
Permutation · 89
PERS · *See* Personal Emergency Response System (PERS)
Personal Area Network (PAN) · 87, 89
Personal Emergency Response System (PERS) · 76, 89
Personal Safety Wearable · *See* Personal Emergency Response System (PERS)
PERT Chart · 89
Peter Principle · 89
Peters, Tom (1942-   ) · 72
P-F Interval · 87
PFEP · 89
Phi Correlation Coefficient · 89
Photocell · 89
Physical Web · 89
Pictograph · 90
Piece Price · 90
Pillar · 45, 90
Pin-Pan-Pon · 90
Pitch · 52, 87, 90
Plan for Every Part (PFEP) · 89, 90
Plan-Do-Check-Act (PDCA) · 7, 8
Planning · 90, 102, 118
Planning and Scheduling · 92
Plant Cost · 59, *See* Inventory Turns
Platform · 90
PLC · *See* Programmable Logic Controller (PLC)
Plug-in · 90
PM · 90
PMTS · *See* Predetermined Motion Time System (PMTS)
PoE · *See* Power over Ethernet (PoE)
Point *Kaizen* · 91
Poisson Distribution · 91

Poka Yoke · 28, 41, 45, 76, 91, 95, 103, 138, 139
Policy Deployment · 91, 137, *See Hoshin Kanri*
Population · 91, 107
Population Correlation Coefficient · 91
Positive Relationship · 91
Positively Skewed Distribution · 91
Power Cycle · 91
Power over Ethernet (PoE) · 91
Power over Wi-Fi (PoWi-Fi) · 91
Power Supply · 92
PPAP · *See* Production Part Approval Process (PPAP)
PPM · 92
Predetermined Motion Time System (PMTS) · 92, 112
Predictive Maintenance (PdM) · 92
Preventive Maintenance (PM) · 92
Principles of Lean · 4, 68, 92, 94
Prioritization Matrix · 92
Proactive Maintenance · 92
Probability · 92
Probability Density Function (PDF) · 31, 88, 92
Probability Tree · 92
Problem Solving · 6, 29, 43, 93, 105, 138, 145, *See* Root Cause Failure Analysis (RCFA), *See* 5 Why, *See* 8D Corrective Action Process
Process · 93
Process Capacity · 93
Process Cycle Efficiency (PCE) · 88, 93
Process Decision Program Chart (PDPC) · 93
Process Evaluation Checklist (PEC) · 93
Process Map · 93
Process Mapping · 93
Process Owner · 93
Process Razing · 93
Product Family · 93
Product Family Matrix · 94
Production Part Approval Process (PPAP) · 94
Production Preparation Process (3P) · 4, 94
Production Smoothing · 94, 106
Program Development (PD) · 88, 94
Programmable Logic Controller (PLC) · 4, 76, 90, 94
Project Evaluation and Review Technique (PERT) Chart · 8, 47, 94
Protocol · 94
Publisher · 26
Pull Production · 4
Pull System/Production · 94, 109
Pulse Width Modulation (PWM) · 94
Push System/Production · 94

P-Value · 87

Q

Q · 95
QAS · 95
QCD · 95, *See* 3 Elements of Demand
QFD · *See* Quality Function Deployment
QMS · *See* Quality Management System (QMS)
QRM · *See* Quick Response Manufacturing (QRM)
QS-9000 · 9, 95
Qualitative Variable · 95
Quality · 30, 86, 95
Quality at the Source (QAS) · 95
Quality Circle · 6, 60, 96, 137
Quality Control · 96
Quality Control Circle · *See* Quality Circle
Quality Function Deployment (QFD) · 96
Quality Management System (QMS) · 95, 96
Quality, Cost, and Delivery (QCD) · 3, *See* 3 Elements of Demand
Quantified Community · 96
Quantified Self · 96, 106
Quantitative Variable · 96
Quartile · 96
Queue Time · 96
Quick Changeover · *See* Smart Changeover
Quick Response Manufacturing (QRM) · 95, 96

R

R · 97
R Chart · 97
Radio Frequency (RF) · 97, 99
Radio Frequency Identification (RFID) · 97, 99
Random Number · *See* Random Variable
Random Sample · 97
Random Variable · 97
Range · 70, 97
Ransomware · 97
Raspberry Pi · 97
RCA · *See* Root Cause Failure Analysis (RCFA)
RCFA · *See* Root Cause Failure Analysis (RCFA)
RCM · 98
Reactive Maintenance · 98, 100
Real-time Production Efficiency (RPE) · 86, 98
Real-Time Production Efficiency (RPE) · 146
Red Tag · 98

159

Reflection · 136, *See Hansei*
Regression · 26, 58, 98, 102
Reliability · 98
Reliability Centered Maintenance (RCM) · 98
Remote Machine Interface (RMI) · 98, 100
Repair · 98
Repeatability · 47, 99
REpresentational State Transfer (REST) · 99
Reproducibility · 47, 99
Resistance · 99
Respect · 3, 4, 45, 68, 99, 135, 136, 138, 139
Respect for Humanity · 138
Respect for Humanity/People · 99
Responsive Maintenance · *See* Reactive Maintenance
REST · *See* See REpresentational State Transfer (REST)
Return on Investment (ROI) · 99, 100
Rework · 99
RF · *See* Radio Frequency (RF)
RFID · *See* Radio Frequency Identification (RFID)
Right-Sized Equipment · 99
Right-Tailed Test · 99
Risk · 99
Risk Priority Number (RPN) · 100
Riso · 139
Risou · *See Riso*
RMI · *See* Remote Machine Interface (RMI)
ROI · *See* Return on Investment (ROI)
Root Cause · 100
Root Cause Analysis (RCA) · *See* Root Cause Failure Analysis (RCFA)
Root Cause Failure Analysis (RCFA) · 4, 5, 7, 44, 93, 100
Router · 100
Run Chart · 100, *See* Control Chart
Run-to-Failure · 98, 100

S

S · 101
SaaS · *See* Software as a Service (SaaS)
Safety · 5
Safety Stock · 40, 101
Saitekika · 139
Sample · 101, 107
Sample Size · 101
Sampling · 17, 101
Sampling Error · 101
Sankey Diagram · 101

SCADA · *See* Supervisory Control and Data Acquisition (SCADA)
Scatter Diagram · 6
Scatter Plot · 102
Schedule Compliance · 102
Scheduling · 90, 102, 118
Scheffé Test · 102
Scientific Management · 102, 111
SCO · *See* Successful Customer Outcome (SCO)
Scrap · 102
Scrum · 102
Seiban · 139
Seicho · 139
Seiketsu · 5, 139
Seiri · 5, 139
Seisan Hoshiki · 139
Seiso · 5, 139
Seiton · 5, 139
Semiconductor · 58, 102
Sensei · 102, 139
Sensor · 18, 102, 115
Sequential Changeover · 103
Serial Communication · 103
Serial Peripheral Interface (SPI) · 103, 107
Serial Set IDentifier · *See* SSID
Series · 103
Server · 19, 103
Service Processes · 123
Setup Reduction · *See* Smart Changeover
Seven Wastes · *See* 7 Wastes
Shadow Board · 103
Shain · 139
Shainin System · 103
Shewhart Cycle · 88, 103
Shewhart, Walter (1891-1967) · 103
Shido (Ryoku) · 139
Shido Sha · 139
Shield · 103
Shine · 5
Shingo, Shigeo (1909-1990) · 103, 139
Ship to Line · 103
Shitsuke · 5, 139
Shojinka · 104, 139
Short · *See* Short Circuit
Short Circuit · 104
Shushinkoyo · 139
Shuukai · 139
SIGFOX · 104
Sigma · 104
Sigma Value · 104
Signal-to-Noise Ratio · 104

SIM · *See* Subscriber Identity Module (SIM)
Simulation Techniques · 104
Single Minute Exchange of Die (SMED) · 103, 105, *See* Smart Changeover
Single Point Lesson (SPL) · *See* One Point Lesson (OPL)
SIPOC · 104
Six Sigma (6σ) · 1, 5, 11, 17, 28, 34, 35, 37, 49, 56, 60, 73, 74, 93, 104, 105, 129, 131, 141, 146
Six Sigma Business Philosophy · 105
Six Sigma Performance · 104, 105
Six Step Problem Solving · 105
Skills Matrix · 105
SKU · 105
Smart · 105
SMART · 105
Smart Building · 105
Smart Changeover · 42, 58, 84, 96, 103, 104, 105, 106
Smart Device · 58, 105, *See* Intelligent Device
Smart Grid · 105
Smart Manufacturing · 4, 56, 106
Smart Meter · 105, 106
Smart Watch · 106
Smartphone · 106
SmartThings · 146
SMED · 42, 58, 84, 104, 105, 139, *See* Smart Changeover
Smith, Bill (1929-1993) · 106
Smoothing · 52, *See* Heijunka, *See* Level Scheduling
SoC · *See* System on a Chip (SoC)
Society of Automotive Engineers · 12
Software as a Service (SaaS) · 101, 106
Sonkei Suri · 139
Sort · 5
Souzouryoku · 139
Spaced Repetition · 106
Spaghetti Chart/Diagram · 106
SPC · *See* Statistical Process Control (SPC)
Special Cause Variation · 28, 107, 122
SPI · *See* Serial Peripheral Interface (SPI)
SQC · *See* Statistical Quality Control (SQC)
SQDC · 107
SSID · 107
Stable Process · 107
Stable/Stability · 107
Standard Deviation · 25, 26, 29, 41, 104, 107, 111, 122
Standard Normal Distribution · 11, 16, 24, 25, 41, 47, 81, 107

Standard Score · 107, 133
Standard Work · 17, 25, 42, 45, 62, 67, 84, 107, 108, 115
Standard Work in Process · 108
Standard Work Practice (SWP) · 108
Standardization · 107
Standardize · 5
Standards · 77
Statistical Process Control (SPC) · 6, 45, 106, 108, 146
Statistical Quality Control (SQC) · 6, 45, 103, 107, 108
Store · 5
Strategic Planning · 53, 108, 137, *See Hoshin Kanri*
Stratification Factors · 108
Structural Variation · *See* Variation
Subscriber · 108
Subscriber Identity Module (SIM) · 104, 108
Successful Customer Outcome (SCO) · 102, 108
Suggestion System · 108
Supermarket · 109
Supervisory Control and Data Acquisition (SCADA) · 102, 109
Supplier · 109
Supply Chain · 26, 109
Sustain · 5, 139
Swim Lane Flowchart · 109
Switch · 109
SWP · *See* Standard Work Practice (SWP)
Synchronous Manufacturing · 109
System of Profound Knowledge · 109
System on a Chip (SoC) · 106, 109

T

T · 111
t Distribution · 111
Takt Time (T_T) · 4, 31, 69, 87, 90, 111, 121
Takt Time Chart · 121, *See* Value Added Chart
Takumi · 139
Tampering · *See* Variation
Task Time Chart · 121, *See* Value Added Chart
Tatakidai · 140
Taylor, Frederick Winslow (1856-1915) · 102, 111
Taylorism · 102, *See* Scientific Management
T-Card · 111
TCO · *See* Total Cost of Ownership (TCO)
TCP/IP · *See* Transmission Control

THE ABIDIAN IIOT/CI DICTIONARY

Protocol/Internet Protocol (TCP/IP)
Team Leader · 112, 136
Tebanare · 112, 140
TEEP · *See* Total Effective Equipment Performance (TEEP)
Teiinsei · 140
Telematics · 112
Tenchou · 140
Tetrachoric Correlation Coefficient · 112
Theory of Constraints (TOC) · 1, 18, 27, 37, 62, 112, 141
Theory X · 112
Theory Y · 112
Theory Z · 113
Therblig Analysis · 48, 92, 112
Thermography · 112
Throughput · 113
Throughput Velocity · 113
TIM WOODS · 113
Time Value Map · 113
Total Cost of Ownership (TCO) · 112, 113
Total Effective Equipment Performance (TEEP) · 112, 113
Total Inventory · *See* Inventory Turns
Total Productive Maintenance (TPM) · 1, 45, 90, 113, 133, 141
Total Productive Manufacturing (TPM) · 1, 2, 45, 90, 113, 133, 141
Total Quality Control (TQC) · *See* Total Quality Management (TQM)
Total Quality Management (TQM) · 30, 114
Toukan · 140
Toyoda · 114, 140
Toyoda, Kiichiro (1894-1952) · 114, 140
Toyoda, Sakichi (1867-1930) · 114, 140
Toyota · 3, 6, 25, 49, 80, 84, 85, 99, 114, 115, 127, 138, 140, 141
Toyota Industries Co., Ltd. · 114
Toyota Production System (TPS) · 84, 85, 99, 114, 127, 138, 140, 143
Toyota Way 2001 · 114
TPM · *See* Total Productive Maintenance (TPM), *See* Total Productive Manufacturing (TPM)
TPS · *See* Toyota Production System (TPS)
TQM · 114
Traceability · 114
Traditional Manufacturing · *See* Mass Production, *See* Batch Operation
Training Within Industry (TWI) · 40, 62, 94, 114, 116
Transceiver · 115

Transducer · 115
Transmission Control Protocol/Internet Protocol (TCP/IP) · 112, 115
Transportation · 28, 106, 113, 115
Travel Time · 115
Traveler Check Sheet · 115
Traveling Turkey · 115
Tree Diagram · 115
Trend · 115
Tribology · 115
TRIZ · 115
True North · 115
Tsukurikomi · 140
T_T · *See* Takt Time (T_T)
t-Test · 111
Turnover · *See* Inventory Turns
TWI · *See* Training Within Industry (TWI)
Two-Bin System · 116
Two-Sided Test · 116
Two-Way ANOVA · 116
Type I Error · 68, 116
Type II Error · 116

U

U · 117
U Chart · 117
UART · *See* Universal Asynchronous Receiver Transmitter (UART)
U-Cell · *See* Celluar Manufacturing
UCL · *See* Upper Control Limit
UEM · *See* Unified Endpoint Management (UEM)
UL or U/L · *See* Uplink (UL or U/L), *See* Upload (UL or U/L)
Underutilized Skills/People/Resources · 113
Uneven · 138
Unified Endpoint Management (UEM) · 117
Uniform Resource Identifier (URI) · 117, 118
Uniform Resource Locator (URL) · 117, 119
Unilateral Tolerance (One-Sided Specification Limit) · 118
Union of Japanese Scientists and Engineers · 63, 118
Unit · 118
Universal Asynchronous Receiver Transmitter (UART) · 117, 118
Universal Serial Bus (USB) · 118, 119
Unix · 21, 69, 118
Unplanned · 118
Unscheduled · 118

Uplink (UL or U/L) · 117, 118
Upload (UL or U/L) · 117, 118
Upper Control Limit (UCL) · 117, 118
Uptime · 118, *See* Availability
URI · *See* Uniform Resource Identifier (URI)
URL · *See* Uniform Resource Locator (URL)
USB · *See* Universal Serial Bus (USB)
User Experience (UX) · 119
User Interface (UI) · 134
Utility Tree/Matrix · 119
Utilization · 119
UX · *See* User Experience (UX)

V

V · 121
V2I · *See* Vehicle-to-Infrastructure (V2I)
V2V · *See* Vehicle-to-Vehicle (V2V)
Validation · *See* Verification
Validity · 121
Value · 121
Value Added · *See* Value Added Activities
Value Added Activities · 21, 40, 80, 121, 122
Value Added Chart · 111, 121
Value Added Time (VAT) · 93, 113, 122
Value Stream · 20, 45, 65, 73, 122
Value Stream Management · 122
Value Stream Manager · 122
Value Stream Mapping (VSM) · 20, 31, 46, 57, 73, 122, 124
Values · 122, *See* Vision
Variable · 122
Variance · 11, 43, 87, 112, 122
Variation · 4, 26, 122
VAT · *See* Value Added Time (VAT)
Vehicle-to-Infrastructure (V2I) · 121, 122
Vehicle-to-Vehicle (V2V) · 121, 122
Venn Diagram · 123
Venture Capital · 30, 123
Verification · 123
Verify · *See* DMADV
Vertical Deployment · 123
Vibration Analysis · 123
Virtual Private Network (VPN) · 123, 124
Virtual Team · 123
Visible Light Communication (VLC) · 69, 123
Vision · 123
Visual Management · 123, 138, *See* Visual Workplace
Visual Workflow · 123
Visual Workplace · 7, 8, 11, 33, 41, 65, 66, 103, 111, 123
Vital Few, Useful Many · 124
VOC · *See* Voice of the Customer
Voice of the Customer (VOC) · 45, 53, 66, 96, 124
Voltage · 124
VPN · *See* Virtual Private Network (VPN)
VSM · *See* Value Stream Mapping (VSM)

W

W · 125
Waiting · 96, 113, 125
WAN · *See* Wide Area Network (WAN)
Waste · 40, 45, 77, 125, 138
Waste Walk · 125
Water Spider · 125
Waterman, Robert · 72
Watson · 125
Wearable · 11, 18, 27, 39, 53, 125
Weave · 125
Web · 126
Web Address · *See* Uniform Resource Locator (URL)
Web Editor · 126
Weibull Curve · *See* Weibull Distribution
Weibull Distribution · 16, 126
Weighted Ranking · 126
Weighted Voting · *See* Weighted Ranking
Weightless · 126
Whisker · 126
White Belt · 126
Wide Area Network (WAN) · 70, 125, 126
Wi-Fi · 3, 5, 8, 19, 51, 126
Wi-Fi Protected Access (WPI) · 126
Wilcoxon-Mann-Whitney Test · 126
Windows · 127
WIP · *See* Work in Process (WIP)
Wisdom · 127
WO · *See* Work Order (WO)
Womack, James P. · 62, 127
Work Cell · 127
Work in Process (WIP) · 69, 84, 108, 127
Work Module · *See* Work Cell
Work Order (WO) · 127
Work Request · 127
Work Team · *See* Natural Team
Workforce · 127
Workout · *See* Kaizen
World Class · 103, 127
WPA · *See* Wi-Fi Protected Access (WPI)

Wristop · 128

X

X · 129
x Bar Chart · 129
X Data/Values · 129
X Dock · 129
XML · 42, 129, *See* Extensible Markup Language (XML)
XMPP · *See* Extensible Messaging and Presence Protocol (XMPP)
XMPP-IoT · 129

Y

Y · 131
Y Data/Values · 131
Yamazume · 140
Yamazumi Chart · 69, 131

Yellow Belt · 126, 131, 146
Yield · 86, 95, 131
Yocto · 131
Yokoten · 131, 140
Yuuryou · 140

Z

Z · 133
z Distribution · 133
z Shift · 133
z Value · 133
Zero Defects · 30, 133
Zero Quality Control (ZQC) · 134
Zero UI · 134
ZigBee · 134
ZQC · 134
z-Score · 133
z-Test · 133
Z-Wave · 133

www.ingramcontent.com/pod-product-compliance
Lightning Source LLC
Chambersburg PA
CBHW071431180526
45170CB00001B/291